Valuing World Heritage Cities

T0227424

With its celebrated World Heritage List, UNESCO steers the global heritage agenda through the definition and redefinition of what constitutes heritage and by offering the highest-level forum for heritage professionalism. While it is the national governments that nominate sites for inclusion in the World Heritage List, and the intergovernmental World Heritage Committee that makes the final decision on inclusion or non-inclusion, it is the International Council on Monuments and Sites (ICOMOS) for cultural heritage that determines whether the necessary level of 'outstanding universal value' is met.

Focusing on the discourses of ICOMOS and their transmission to the local context, this book is the first in-depth historical analysis of the construction of heritage value in the context of cities illustrated through a case study of Old Rauma in Finland. The book contributes to the understanding of the discursive and constructed nature of World Heritage values as opposed to intrinsic values, critically scrutinizes the role of ICOMOS in making valuations concerning urban heritage, and sheds light on the interactions and tensions of universal and local (urban) perspectives in the practice of heritage valuation.

Valuing World Heritage Cities is the first in-depth historical analysis of the construction of heritage value in the context of cities in the transnational discourses of heritage. This unique and timely contribution will be of interest to scholars and students working in Heritage Studies, Cultural Geography, Urban Studies and Tourism.

Tanja Vahtikari is Postdoctoral Research Fellow at the Finnish Centre of Excellence in Historical Research at the University of Tampere, Finland.

Routledge Cultural Heritage and Tourism Series
Series editor: Dallen J. Timothy, Arizona State University, USA

The *Routledge Cultural Heritage and Tourism Series* offers an interdisciplinary social science forum for original, innovative and cutting-edge research about all aspects of cultural heritage-based tourism. This series encourages new and theoretical perspectives and showcases groundbreaking work that reflects the dynamism and vibrancy of heritage, tourism and cultural studies. It aims to foster discussions about both tangible and intangible heritages, and all of their management, conservation, interpretation, political, conflict, consumption and identity challenges, opportunities and implications. This series interprets heritage broadly and caters to the needs of upper-level students, academic researchers and policymakers.

Valuing World Heritage Cities
Tanja Vahtikari

Waterways and the Cultural Landscape
Edited by Francesco Vallerani and Francesco Visentin

Heritage of Death: Landscapes, Sentiment and Practice
Edited by Mattias Frihammar and Helaine Silverman

Industrial Heritage and Regional Identities
Edited by Christian Wicke, Stefan Berger and Jana Golombek

Valuing World Heritage Cities

Tanja Vahtikari

LONDON AND NEW YORK

First published 2017 by Routledge

2 Park Square, Milton Park, Abingdon, Oxfordshire OX14 4RN
711 Third Avenue, New York, NY 10017

Routledge is an imprint of the Taylor & Francis Group, an informa business

First issued in paperback 2018

British Library Cataloguing in Publication Data
A catalogue record for this book is available from the British Library

Library of Congress Cataloging in Publication Data
A catalog record has been requested for this book

ISBN: 978-1-4724-6102-5 (hbk)
ISBN: 978-1-138-54690-5 (pbk)

Typeset in Times New Roman
by diacriTech, Chennai

To Okko and Iisa, my World and my Heritage

Contents

Figures

Acknowledgments

There are many individuals and organizations that I am indebted to at the moment of ending this project – the list that follows represents only a small selection. First of all, I would like to thank Marjatta Hietala and Marjaana Niemi for introducing me to the world of urban history, for sharing expertise and for encouragement, support and enthusiasm that never failed. My sincerest gratitude also goes to those who read earlier versions of the manuscript, including Luda Klusáková, Taina Syrjämaa, Rebecca Madgin, Marc Schalenberg, Derek J. Oddy and Michael Berry. I also both enjoyed and benefitted greatly from discussions with Margaretha Ehrström, Håkan Forsell, Maunu Häyrynen, Jaroslav Ira, Mervi Kaarninen, Sari Katajala-Peltomaa, Linda Kovarova, Silja Laine, Henrik Mattjus, Anna Nurmi-Nielsen, Sari Pasto, Kalle Saarinen, Dieter Schott, Katri Sieberg, Mervi Tammi, Timo Vilén and Eliisa Vähä. In addition, I have had the opportunity to discuss my work in several conferences and workshops at home and abroad that provided inspiration. Of those, I would especially like to mention the European Association for Urban History, with its conferences in Prague in 2012 and in Lisbon in 2014, and the roundtable "National Identities and World Heritage," convened by Tamara von Kessel and Pim den Boer in association with the twenty-second International Congress of Historical Sciences, held in Jinan, China, in 2015. At Routledge, I am grateful to Faye Leerink and Priscilla Corbett for their invaluable guidance in the various steps towards publication. I would also like to sincerely thank Series Editor Dallen J. Timothy, who welcomed my book to be a part of the new Routledge Cultural Heritage and Tourism Series.

While finalizing the manuscript I have been privileged to work at the Finnish Centre of Excellence in Historical Research funded by the Academy of Finland. I would like to thank the director of the Centre, Pertti Haapala, for that opportunity. I am equally grateful to the Kone Foundation research project, "Experts, Communities and the Negotiation of the Experience of Modernity," led by Marjaana Niemi, for funding. The research community in the discipline of history at the School of Social Sciences and Humanities of the University of Tampere is unequalled – thank you all for listening and sharing! A special word also goes to the helpful staff in the UNESCO Archives and the World Heritage Centre in Paris, in the Archives of the National Board of Antiquities in Helsinki and in the Rauma City Archives. In connection with archival material I should mention that UNESCO has done a wonderful job of digitalizing the majority of the World Heritage–related materials, thus many sources are now just a click away from one's home computer. Here I would also like to acknowledge the generous permissions I have received to use photographs. It is obvious, in addition, that all my interviewees in Rauma and at the National Board of Antiquities deserve very special gratitude. I would also like to express my sincerest gratitude to Carl Wieck, who not only patiently corrected the faults of my English language but also shared invaluable advice throughout the writing process.

My very special thanks go to my dear friends and two families, Vahtikari and Berry. My parents, Timo and Marjatta, and my sister Hanna have supported me all the way. Thank you so much for that! Touko, Okko and Iisa – this book is for you, with love.

1 Introduction

With its celebrated *Convention Concerning the Protection of World Cultural and Natural Heritage* (1972)[1] UNESCO steers the global heritage agenda by defining and redefining what constitutes heritage and by offering a high-level forum for heritage professionalism. World Heritage sites, as understood by UNESCO, are places that have "outstanding universal value." While it is the national governments that nominate sites for inclusion in the World Heritage List, and while it is the intergovernmental World Heritage Committee which makes the final decision on inclusion or non-inclusion, it is the International Council on Monuments and Sites (ICOMOS) for cultural heritage that examines whether the level of outstanding universal value is met. During this process of establishing value, a place that had previously been recognized as locally and nationally significant is given an additional layer of meaning. When the official statement of outstanding universal value has been decided on, it becomes critical as to how this abstract notion of value is interpreted and used locally, and questions regarding World Heritage become particularly critical at any time of dissonance. For instance, World Heritage has produced conflicts in connection with many cities, including the Finnish city of Rauma, where plans to build a large shopping center on the buffer zone of Old Rauma, a World Heritage site since 1991, became openly debated in the mid-2000s.

This book explores the time-bound and multi-layered construction of outstanding universal value in the context of World Heritage inscribed cities during the nearly forty years of implementation of the World Heritage Convention (1972–2011). By discussing what has been valued as/in a World Heritage city, and why, it provides the first in-depth historical analysis of the construction of urban heritage values in the transnational discourses on heritage. While not entirely distancing the national level from the analysis, with regard to the various levels of heritage, this book focuses on the *international* and the *local*. The global aspect is approached primarily through analyzing the articulation of outstanding universal value by ICOMOS in its statements written regarding the qualities of World Heritage nominated cities. The local is represented by Old Rauma.

This book has three main objectives, all offering fresh viewpoints concerning the scholarship. First, it examines the agency of ICOMOS in the World Heritage valuation procedures, something which has been largely left unproblematized in earlier studies. The statements compiled by ICOMOS act as intermediaries between nationalist discourses on cultural significance articulated in the World Heritage nominations by states and the final decisions made by the World Heritage Committee. As such, they play a central role in the World Heritage valuation and decision-making framework. The statements by ICOMOS provide a classic case for the assessment of cultural significance and the assigning of meaning to heritage. As I will argue throughout, this is also how they should be treated, rather than as value-free "evaluations."

Second, the book contributes to the understanding of the constructed nature of World Heritage values as opposed to intrinsic values, and to the discussion first initiated by Laurajane Smith concerning the hegemony of an internationally working "authorized heritage discourse," privileging "monumentality and grand scale, innate artefact/site significance tied to time depth, scientific/aesthetic expert judgement, social consensus and nation building."[2] I will ask what kind of social and cultural messages have been conveyed by ICOMOS in its articulation of outstanding universal value. Despite the centrality of values to all heritage work, there exist very few studies that elaborate on the complex processes of World Heritage valuation, and none that has a similar focus to this book with respect to the valuation practices of ICOMOS and the transmission of the concept of outstanding universal value to the local context.

An important contribution to today's scholarship is the recent work by Sophia Labadi, which offers theoretical perspectives on outstanding universal value, and elaborates on the question of how states, in their World Heritage nomination dossiers, have understood outstanding universal value in regard to postnational histories, cultural diversity, sustainable development and authenticity.[3] Along with Labadi, I share an interest in investigating the practical articulation of value as part of the World Heritage–related processes, but we approach the theme from two different angles – she from the point of view of States Parties and UNESCO, and I from the perspective of ICOMOS and the local community – and in relation to different categories of heritage – she focuses on religious and industrial heritage, and I on urban heritage. Labadi identifies values related to the history and development of properties, their architectural and aesthetic descriptions, and references to men from the middle and upper classes as dominant in the framework of national nomination dossiers.[4] This offers a good point of reflection from which to explore whether or not ICOMOS, according to its own definitions, has departed from the articulation of values by states, which often use the Convention as a nationalist instrument.[5] Equally, it enables a discussion of cities in comparison to other categories of heritage, evoking the question of whether the "urban" has formed a distinctive category as part of the World Heritage valuation process.

Since the 1970s, the perception of the concept of cultural heritage has undergone significant widening in professional conservationist and academic circles, as well as among the broader public.[6] This has involved the more inclusive perspective as to what is to be considered a legitimate part of cultural heritage, as well as to increased reflection regarding (a broadening spectrum of) values, and, in the field of urban heritage and within the integrated conservation approach, the treatment of continuity, re-utilization and adaptation to the contemporary life of historic urban areas as part of their protection.[7] As will be discussed in Chapter 3, these widening conceptualizations of heritage in society have also had a direct impact on the implementation of the World Heritage Convention. This raises the question of how ICOMOS, when assigning outstanding universal value, has responded to this evolving understanding. Has ICOMOS been successful in promoting pluralization of values and narratives? These questions are particularly relevant, as one of the key objectives of the World Heritage organization today is being in the forefront of the development of standards and practices in urban conservation.[8]

Sometimes the statements that ICOMOS makes are negative, when it is unable to establish the outstanding universal value which the States Parties have proposed in their nomination dossiers. These negative statements, together with the rejections by the World Heritage Committee, represent a theme that has been rarely discussed either by scholars or by the official World Heritage organization.[9] UNESCO and ICOMOS have been reluctant to reflect on the matter out of discretion towards those states whose nominations have been rejected. From the point of view of the World Heritage community an ideal situation would be for there to be no rejections whatsoever, but for non-suitable candidates to not even be nominated at all by the states. Along the lines of this consensual approach, one of the objectives of an action plan proposed by ICOMOS in 2005, to allow States Parties to contribute to the development of the World Heritage List, was "to optimize the success of World Heritage nominations."[10] In this book ICOMOS' negative statements concerning outstanding universal value of cities will be discussed along with its positive ones. An interesting question is how the discursive line has been drawn between successful and unsuccessful World Heritage nominations and between outstanding universal value versus 'other' value.

The third objective of the book is to shed light on the interactions and tensions of global and local perspectives and processes in the practice of heritage valuation. Whilst there are several studies that examine the globalizing effect of the World Heritage Convention within a local context,[11] these discussions do not take the concept of outstanding universal value as their point of departure. I will ask how outstanding universal value gets articulated and practiced locally subject to diverse interests within the context of the individual World Heritage city, Old Rauma. I argue that the construction of outstanding universal value does not end at the point of inscription on the World Heritage List: it continues after the designation at the local level between different groups (local but also translocal), and with regard to locally bound and historically constructed concepts of place and heritage. In this part of the research I emphasize the nature of heritage as a process which builds local identities and which involves actions and debates. Therefore, it seems pertinent to ask how the earlier meanings and practices of a place have become re-defined in the context of the World Heritage site. I will also ask what kind of value the outstanding universal value has represented when used locally, especially at a time of heritage dissonance, and which elements of the international discourse have been transmitted to the local level, and how and why. To what extent has outstanding universal value become part of the local understanding of a place and heritage? When in conflict, which levels of heritage have taken priority?

Heritage: multi-scale social and cultural construction

All environments are equally historical. Historians may select some of them according to the very narrow limits of their discipline and call them "historic," but the difference between "historical" and "historic" is an historian.[12]

The above distinction by Peter Howard between historical and historic environments summarizes many key aspects related to heritage,[13] which forms the key conceptual and theoretical point of departure in this research: its selectivity, its 'presentness' and time- and context-bound nature, and the central role played by experts in its definition and valuation. No past artefact becomes heritage self-evidently: while some places get remembered, produced and managed as heritage, others get demolished, or remain – consciously or unconsciously – hidden and forgotten. As pointed out by Jennifer Jordan, "[f]ew places bear the traces of their past unaided."[14] Selection gives heritage its ultimate worth. In the burgeoning literature on heritage which has appeared over the past three decades there has been wide acceptance of the view that heritage is constructed through various cultural and social processes.[15] This view is also shared in this research. Heritage is more than the material remains of the historical process, or the act of conservation[16] of these remains. Heritage is not a thing but a meaning, a process and a relationship with the past. As such it has a transformative nature.[17]

The gaze of heritage is directed towards the past but is always interpreted from the current perspective, for present and future purposes, and is infused with the concerns and uses of the present. This is the case whether we identify heritage primarily as professional conservation, an inheritance and an intergenerational relationship in the processes of collective memory and identity, as national patrimony aiming to situate the nation in a significant historical sequence, or as an economic resource for the marketing of place and for tourism. Equally, heritage formation is, for instance, closely related to issues of accumulation of knowledge or present land ownership and land use.[18] Each period and each culture defines heritage from its own perspective. At the beginning of the twenty-first century "heritage has become a quasi human right,"[19] and "a strategy for the future," treated hand in hand with strategies related to human rights, sustainable development, climate change, and tourism.[20]

In light of its 'presentness,' it is not surprising that heritage often serves as a dissonant, fragmenting and exclusive resource rather than a unifying and inclusive one. It is contested along several axes: public/private, cultural/economic, temporal and spatial.[21] The inherent dissonance of heritage also relates to its creation by interpretation, to its association with values and valuation, and to the fact that it is always valued, interpreted and 'owned' by someone. Consequently, there are power issues involved, and one further arena of contestation can be identified as existing between expert and non-expert values. Whilst wider community participation in the cultural heritage field is increasingly encouraged today, the heritage truths are still often produced under the control of heritage experts and expert institutions.[22] Clearly, World Heritage cities represent no less contested resources than any other form of heritage. Quite the contrary, it may be argued that World Heritage is especially laden in this area.

Central to the idea of heritage is 'value.'[23] No society preserves something it does not value. Heritage sites are identified, listed, conserved, managed and interpreted according to the values they are seen to represent. The challenge in articulating heritage values stems from the fact that many diverse values exist simultaneously, in coexistence or in competition with each other, and that values change over time and in different societal contexts. The official professional discourses on heritage have tended to treat heritage values as something intrinsic to a particular place or a building.[24] However, if we take into account the above-discussed recent theoretical orientation in heritage studies, heritage may be seen as a continuous cultural process in which social and cultural meanings and values are created, negotiated and transmitted, rather than a static object of intrinsic value. Heritage values are socially and spatially constructed, extrinsic and contingent.[25] If one shares this thought, as I do in this research, all the actions involving heritage places, including their identification, listing, conservation, management and interpretation, should be seen as a forming part of heritage and not as purely external treatments of it.[26] Nevertheless, it should be noted that the constructionist approach to heritage values is not without conceptual dissonances. On the one hand, considering all values to be equally true can potentially result in a situation where there is no possibility of any shared narrative or norms of good practice.[27] On the other hand, taking a fully relativist stand may lead to the conclusion that all ways of treating heritage are equally acceptable, including its purposeful destruction. The widespread condemnation by the international community of the recent destruction of heritage properties, such as the Buddhas of Bamiyan in Afghanistan, suggests that there could exist a certain consensus with respect to cultural heritage-related basic ethical values.[28] In an endeavor to take into account both universal and relative values, Sophia Labadi introduces the concept of "reiterative universalism." Key to this idea is the notion that universal concepts of World Heritage are actively translated into culture-specific frames of reference.[29]

It is widely held that the meaning making involving cultural heritage takes place on various spatial scales and levels of practice – sometimes overlapping and complementary, sometimes conflicting and contested – ranging from individual, neighborhood and local levels, to regional, national and even global scales. Cultural heritage is a key instrument in the construction of identities at various levels, and place identities are central in the creation and support of respective spatio-political entities.[30] Furthermore, cultural heritage is valued, managed and presented at various hierarchically established locations, extending from the local to the international. The very idea of significance assessment in relation to cultural heritage presumes difference in the amount and level of value.

Several authors have shown the dominancy of the national over other scales in reference to the inextricable historical link between nation states, nationalism and national heritage projects,[31] even though what has been considered as national heritage has had a transnational dimension from the the late eighteenth century onwards.[32] The priority of the national claim over other scales of heritage has been further supported by the fact that the majority of the institutions and the major portion of the legislation protective of cultural heritage have been created within the framework and for the purposes of nation states. What is treated as national heritage fragments into several sub-categories, such as heritages of regions and cities, which support local identities. The globalization of heritage, which has intensified since the 1960s, represents the other side of the spatial spectrum. Over the past few decades several trends have supported the idea of global heritage: increasing global environmental concern; the establishment of international institutions in the field of cultural heritage; international tourism, for which cultural heritage forms a key resource; and the unfortunate role given to destruction of heritage as part of international conflicts. Hitherto, the World Heritage Convention has been the most advanced institutionalized articulation of the idea of a common global heritage. While it is important to scrutinize the international-level construction of World Heritage, as will also be done here, it is equally important to note the multi-level nature of the phenomenon: World Heritage gets articulated and re-articulated at all levels of the heritage scale from the global to the local.

As several authors convincingly show, heritage is a discursive construction shaped by specific circumstances – a discourse here referring to something that is both reflective and constitutive of social practices.[33] In other words, heritage and its meaning are "constructed *within*, not above or outside representation."[34] Whilst acknowledging the existence of several simultaneous and competing heritage discourses, Laurajane Smith argues that Western conservation practice since the late nineteenth century has developed towards a dominant position as a universalizing discourse of our own time. This "authorized heritage discourse" (AHD), located in between "power/knowledge claims of technical and aesthetic experts" and state cultural agencies, is a self-referential discourse – its operation influences the way we talk about and otherwise represent heritage.[35] Smith discusses the World Heritage Convention as one of the key international legitimizing institutions of AHD. Her discussion, however, is focused mainly on the issue of the Eurocentrism of the World Heritage List and the existing listing criteria.[36] One aim of this book is to discuss the operation and extent of AHD within the framework of ICOMOS statements, and in relation to the politics of representation of World Heritage cities in particular – what has been privileged by this discourse, how stable or mutable it has been over the years, whether it has been challenged, and if so, in what ways and to what extent.

It needs to be emphasized, however, that the discursive nature of cultural heritage outlined above in no way aims to deny the material aspects of heritage or its association with specific places. Rather, the tangible, material and place-bound side of heritage should be seen to co-exist with the discursive side.[37] Therefore, a heritage place can be understood as an interchange between the discursive representation, the physical space and the socially lived place.[38] This idea of the importance of the place as a "politicized, culturally relative, historically specific, local and multiple construction"[39] is inherent in the discussion found in Chapter 6.

World Heritage cities: accepted and rejected

The basic distinction drawn by UNESCO among World Heritage sites is the one between cultural and natural heritage.[40] The composition of the World Heritage List, like any representation of heritage, is a subject of changing perceptions in society and a result of selection. During past decades, one of the major emphases in the heritage field has been placed on urban heritage, also within the context of World Heritage. For the purposes of this research, a World Heritage city is understood to be an inhabited urban area which sustains everyday activities and supports a living urban fabric. To convert Leonardo Benevolo's words into the context of World Heritage cities, they are "places in which it is possible to access and absorb cultural values in every day activities with a consideration that is not temporarily limited to short term visits."[41] Based on the above considerations I have identified altogether 187 World Heritage sites, inscribed between the years 1978 and 2011, as World Heritage cities (Appendix 1). Most of the identified sites are comprised of (sections of) historic city centers, but a few cultural landscapes, the outstanding universal value of which was defined primarily in reference to urban features, were included in the data.[42] Here it should be pointed out that drawing a line between World Heritage cities and other World Heritage sites (archaeological urban sites, villages, monumental squares or industrial complexes within the city without including the surrounding urban fabric) remains an ambiguous exercise. Even the definitions in effect within the World Heritage organization have often varied depending on who makes them and for what purpose.[43]

Based on the same qualification criteria as with inscribed World Heritage cities, I singled out altogether 41 cases of rejected nominations of historical cities between the years 1978 and 2011 (Appendix 2). This means a little over one rejection per year on average, even though in reality the denials, like the inscriptions, have been more unevenly distributed. As Figure 1.1 shows, the low year of urban as well as other inscriptions was 1989, with only four cultural sites and no urban sites included in the World Heritage List. That same year was also the highest year for rejections (altogether six). Another peak year was 2000, which in the implementation of the Convention marked a year when ICOMOS evaluated the highest ever number of cultural nominations (70, including three extensions).[44] There were six rejections of urban nominations that year, all of which were actually withdrawals by the relevant State Party after a negative statement issued by ICOMOS. Here it should be emphasized that the list provided in Appendix 2 includes both the nominations that were officially rejected by the World Heritage Committee in its yearly meetings, and those nominations whose rejection was recommended either by ICOMOS in its evaluation and/or by the Bureau of the World Heritage Committee, but which the State Party decided to withdraw before the actual Committee meeting. Withdrawing nominations has been a regularly used

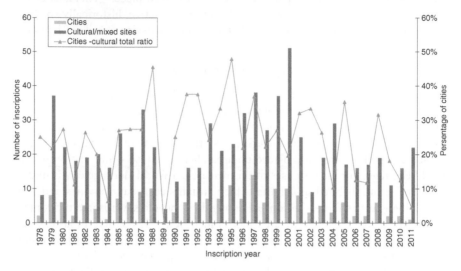

Figure 1.1 Yearly inscriptions of cities on the World Heritage List in relation to all
       inscribed cultural and mixed sites.

practice among States Parties, enabling them, in a situation of pending rejection
of a site, to avoid the embarrassment and disappointment caused by this action.

Forty-one rejections until 2011, including nominations withdrawn by states,
indicate a high acceptance percentage of national nominations. What nevertheless needs to be remembered here is that in addition to the official category of
rejection, there are two other categories of non-inscription in the World Heritage
designation system: deferring and referring back of the nomination to the State
Party. And in practice a referral back or a deferral of a nomination has sometimes
had the same effect as an official rejection – the end of an inscription process.
This happened, for example, in the case of Spanish Town which was nominated by
Jamaica and evaluated by ICOMOS in 1988. Its deferred status "until measures
have been adopted to assure the conservation, rehabilitation and restoration of the
historic centre" is still in effect,[45] but Spanish Town no longer is included in the
up-to-date tentative list of Jamaica.[46] To provide another example, the application
of the Early Medieval Architectural Complex and Town of Panauti (Nepal) was
referred back to the State Party in 1998 pending the receipt of more information
about the legal protection and management of the site.[47] Apparently no such information was ever received and the Town of Panauti, as with many other deferred/
referred back nominations, disappeared from the agenda during the inscription
process. As ICOMOS indicated in 2008, the registered number of cultural sites
in 2007 amounted to 1,265, whereas at the same time the total number of cultural sites actually inscribed on the List was 851. This means that about a third
of cultural nominations had been subject to non-inscription. A smaller sample of
cultural nominations from 2003–2007 indicates that around 35 percent of non-
inscriptions had been officially rejected.[48] This practice is concomitant with the

World Heritage organization's overall consensual approach: when unfitting nominations get 'buried' in the process, the more dissonant act of rejection does not need to be carried out so often.

Data and methods

The two-scale research approach applied in this study has meant that a great variety of data has been utilized. In order to map the general World Heritage discourse and its evolvement over the years, an extensive archival analysis of UNESCO documents has been carried out, with a particular focus on documents relevant from the perspective of urban heritage. These documents include the World Heritage Committee, Bureau and General Assembly minutes, a number of the working documents from these meetings, as well as reports compiled by ICOMOS and the World Heritage Centre. Furthermore, I have followed the yearly inclusions of cities in the World Heritage List.

Considering the focus of the research on the ICOMOS valuation practice, the most informative source has been the evaluation documents compiled by this organization, both those recommending inscription and those recommending rejection.[49] There has been a significant widening in terms of both the content and length of these documents over the years. From 1979 onwards, the evaluation documents included a brief identification of the property, background and justification, a reference to relevant criteria and the ICOMOS recommendation – as to whether the nominated site should be included, deferred or rejected. In 1992 the format of evaluation documents was changed and more categories were added; "authenticity," "management and protection" and "comparison with other sites" appearing for the first time as separate subtitles. "Involvement of the local communities" was added as an independent section to ICOMOS evaluations starting in 2007. At present the evaluation documents, after giving the basic data concerning the nomination, describe the history and development of the nominated site; reflect on the outstanding universal value, integrity and authenticity; and discuss the factors affecting the site (most importantly development pressures), its protection, conservation, management and future monitoring. Today the ICOMOS evaluations are written more in the form of a joint dialogue together with the nomination dossiers of the states. In this situation there is more explicit commentary than previously on various issues that the national nominations raise.

The restricted scope of the earlier evaluation documents, especially those which appeared prior to 1992, causes certain limitations concerning their analysis. It should be noted, however, that the explicit requirements after 1992 were often implicitly present in the earlier reports. Furthermore, the reference, or lack of reference, to certain aspects at certain times ought to be viewed as descriptive of the valuations of that particular period. The question of authenticity, for example, was seriously discussed in the context of post-Second World War conservation; the requirement for meeting "the test of authenticity" for cultural heritage was also included in the World Heritage valuation system. During the 1970s and 1980s, however, the question lost out to other issues (especially those related to scientific development in the area of restoration) considered by international heritage professionals. The authenticity question was revived only in the early 1990s.[50] This was also reflected in ICOMOS evaluations of World Heritage nominations.

After defining the group of World Heritage cities to be studied, including those not inscribed, a qualitative textual analysis of central themes and values referred to in their respective evaluation documents was conducted. The ICOMOS evaluations were read as a large body of text, categorizing it in relation to various themes along the axes of monumental versus non-monumental, historical and aesthetic values versus socio-cultural values, authentic versus inauthentic, stability versus change, urban continuity versus discontinuity, unity versus diversity, tangibility versus intangibility, contextual versus non-contextual, European versus non-European, and national versus global/transnational/multicultural/local. I examined to what extent these different themes and sets of values were referred to, and, more importantly, in what manner they were promoted. The themes were developed on the basis of close acquaintance with the official World Heritage minutes, the main international charters and recommendations in the field of urban heritage, a number of the existing heritage value typologies (see Chapter 2), and research literature on critical heritage studies; they were, in addition, further refined during the process of reading the ICOMOS evaluations.

My analysis of the ICOMOS evaluation texts has also been influenced by the Critical Discourse Analysis, and especially by the application of this theoretical and methodological approach in heritage studies by Emma Waterton, Laurajane Smith and Gary Campbell.[51] Critical Discourse Analysis is concerned with unpacking the seemingly natural and common-sense discursive formations, and brings awareness that even apparently neutral and informative texts and images are "just as implicated in the process of conveying power and status as more explicitly ideological" ones.[52] However, Critical Discourse Analysis distances itself from the extreme position that holds that everything can be reduced to the level of discourse.[53] Waterton, Smith and Campbell use Critical Discourse Analysis to review one key text in the international professional heritage field, the *Burra Charter: The Australia ICOMOS Charter for Places of Cultural Significance* (hereafter referred to as the Burra Charter), originally written in 1979 and since then modified several times, the latest modification occurring in 2013. The authors discuss the overall textual organization of the 1999 Burra Charter version and try to "isolate specific semantic and grammatical instances in which important discursive work is done."[54] They conclude that despite sincere attempts towards more inclusive practices in the Charter's revisions, "the discursive construction of the Burra Charter effectively undermines these innovations."[55]

Because of the large quantity of data used in this book (the over 200 ICOMOS evaluation documents alone comprise a seriously large body of text), the same kind of close reading of text as practiced by Waterton, Smith and Campbell with regard to the Burra Charter has not been possible. My work has nevertheless benefitted from several concepts introduced by the Critical Discourse Analysis approach. According to Norman Fairclough, specific principles of recontextualization "underlie differences between the ways in which a particular type of social event is represented." These principles include: "Presence" (which elements of events are present/absent, prominent/backgrounded); "Abstraction" (what is the degree of generalization from concrete events); "Arrangement" (how events are ordered); and "Additions" (what is added to representation of events – explanations, legitimations). Fairclough also notes the choice in the representation of social actors (inclusion/exclusion; activated/passivated; personal/impersonal).[56] Other useful concepts are intertextuality and dialogicality. Intertextuality refers to framing a text in relation to other texts (explicitly and more implicitly), whereas dialogicality points to how different texts engage in a dialogue to varying degrees.[57]

For the local, Rauma part, the variety of data applied has been diverse, ranging from planning documents and newspaper articles to 28 thematic, semi-structured interviews (averaging 1.5 hours), which I conducted mainly in 2002 and 2003 with 27 interviewees, representing various groups involved in giving a meaning to Old Rauma as broadly as possible, and selected by utilizing a purposive and snowball sampling technique. These groups include inhabitants, shopkeepers and house owners of Old Rauma, city conservation and town planning authorities, members of the Old Rauma Society, local politicians and those national-level conservation authorities who had been involved with the conservation and the World Heritage nomination of Old Rauma. The underlying notion was that every interviewee would speak for her/himself as an individual, and also communicate as a representative of a broader group.[58] The interviews dealt with the interviewees' understanding of Old Rauma's outstanding universal value but also with many other issues related to conservation and the local implementation of the World Heritage Convention. They provided the core of my understanding of how local groups in Rauma have interpreted the (outstanding universal) value of Old Rauma. Furthermore, they provided a firm basis for analyzing the World Heritage–related meanings and the shopping center debate in the local continuum.

Chapter overview

The chapters in this book bring together the broader themes of the role of experts as part of the processes of heritage, the interaction of transnational and local in the heritage practice, and heritage as a selective social construction and a discursive practice. The key rationale and motivation behind the entire World Heritage effort is the concept of outstanding universal value. The meaning of outstanding universal value has been (re-)constructed over time and through several interrelated processes and by various stakeholders when putting the World Heritage Convention into practice. These include the nominations (and non-nominations) to the World Heritage List and tentative lists by the States Parties to the Convention; the case-by-case (e)valuations, compiled by ICOMOS for cultural heritage and IUCN (the International Union for Conservation of Nature) for natural heritage; the regularly revised criteria used in this assessment; the numerous comparative analyses by the advisory organizations dealing with different types of heritage; and the actual decisions – inscriptions, deferrals and rejections – by the World Heritage Committee. In addition to this case-by-case construction, the World Heritage Committee and its advisory bodies have regularly provided more general explanations of the concept of the outstanding universal value. After placing the World Heritage Convention in the context of the postwar heritage concern, the following chapter will discuss the main trends visible in these conceptualizations over time, especially the meanings of 'universal' and 'intrinsic value' as part of the definition of outstanding universal value. The chapter then moves on to analyze the values articulated in the World Heritage Convention and its Operational Guidelines in regard to recent heritage value typologies. Chapter 2 ends by exploring the expert and mediator role of ICOMOS in relation to national governments and the World Heritage Committee in the processes of articulating outstanding universal value. The World Heritage truths continue to be produced under the control of experts.

Since the early 1990s a forceful rhetoric of change and reinterpretation has been a distinctive feature of the official World Heritage discourses, framed within the context of broader debates in the professional heritage field and in the wider society. Building on earlier research, Chapter 3 addresses five elements of this reorientation: launching the Global Strategy for a more representative World Heritage List; introducing a new category of cultural landscapes for World Heritage nominations; widening the Western-based concept of authenticity; bringing the notion of intangible heritage into the debates on World Heritage; and assigning a larger role to local communities in defining World Heritage. Elaborating on these five elements is important because the following chapters consider in what forms, and especially to what extent, this widening of scope has really taken place in the context of valuing World Heritage cities. Chapter 3 closes by looking into the role that the "urban" has played as part of the reassessment of what constitutes World Heritage.

Chapters 4 and 5 shift the focus to World Heritage cities, and to the manner in which ICOMOS has textually represented these cities and their outstanding universal value. The two chapters are organized around the questions "Which pasts and whose histories?" and "What urban futures?" Historical value, the capacity of a place to convey a relationship to the past, lies within the very core of heritage valuation. By analyzing the different historical values associated with urban World Heritage, Chapter 4 explores ICOMOS' building of chronologies and dominant and marginal themes, the negotiation between national and transnational meanings, treatment of negative heritage and conflicting and alternative narratives of history, and the naming (and taming) of cities as World Heritage sites. These issues will be approached by exploring what kinds of cities, over time, have been inscribed on the World Heritage List, in what ways ICOMOS has applied the qualification criteria for inscription in relation to the notion of historical value, and, most importantly, what kinds of representations of history have been created as part of ICOMOS' description of cities. In addition to historical values, World Heritage cities are associated with many other values. These other values, especially monumental, aesthetic, contextual/environmental and social/community values together with the notions of authenticity, reconstruction and change, key to urban conservation, are addressed in Chapter 5. The subtitle of the chapter, "What urban futures?" suggests that the articulation of these values has consequences for the futures of the cities designated as World Heritage. There is very little doubt that a broadening of values and themes has taken place in the assessment of cultural heritage, or cities for that matter, in the context of World Heritage over the years, especially since the mid-1990s. It is interesting, however, to consider in what forms, and especially to what extent, this widening of scope has really taken place in practice. Whilst demonstrating a significant conceptual widening in the context of World Heritage cities over a relatively short period of time, Chapter 5 illustrates the simultaneous difficulty involved in enabling many of the noble ideas to go beyond mere rhetoric.

Chapter 6 brings the discussion down to the local level, and to the question of how the concept of outstanding universal value operates locally. It starts by enquiring into the history of Old Rauma as heritage and as a historic city prior to its official designation in 1991 as belonging to World Heritage. It then discusses the establishment of World Heritage value for Old Rauma by comparing the value statement articulated by the Finnish State Party with the one produced by ICOMOS. The remaining part of the chapter explores the operation of outstanding universal value locally after 1991, with particular focus on the debate concerning the construction of a shopping center in the World Heritage buffer zone of Old Rauma between 2003 and 2006, something which finds parallels in many other World Heritage–designated cities. It will be shown, for instance, how the open-ended concept of outstanding universal value was used by various groups to promote their own particular visions in negotiations over place and heritage meanings.

Notes

1 Hereafter, this legal instrument is referred to in the text either as the World Heritage Convention or as the Convention.
2 Laurajane Smith, *Uses of Heritage* (New York: Routledge, 2006), 11.
3 Sophia Labadi, *UNESCO, Cultural Heritage, and Outstanding Universal Value. Value-based Analyses of the World Heritage and Intangible Cultural Heritage Conventions* (Lanham: AltaMira Press, 2013); Sophia Labadi, "Representations of the Nation and Cultural Diversity in Discourses on World Heritage," *Journal of Social Archaeology* 7: 2 (2007): 147–170. For UNESCO's perspective on outstanding universal value see also Sarah M. Titchen, *On the Construction of Outstanding Universal Value: UNESCO s World Heritage Convention and the Identification and Assessment of Cultural Places for Inclusion in the World Heritage List.* Unpublished PhD. diss., Australian National University, 1995.
4 Labadi, "Representations of the Nation," 158.
5 Labadi, *UNESCO*, 149.
6 See, for example, Raphael Samuel, *Theatres of Memory. Vol. 1. Past and present in contemporary culture* (London: Verso, 1994); Françoise Choay, *The Invention of the Historic Monument* (Cambridge: Cambridge University Press, 2001); Jukka Jokilehto, *A History of Architectural Conservation* (Oxford: ICCROM, 1999); concerning archaeology see Denis Byrne, "Western hegemony in archaeological heritage management." *History and Anthropology* 5: 2 (1991): 269–276; concerning anthropology see Anna Tsing, "The Global Situation." *Cultural Anthropology* 15: 3 (2000): 339; concerning the emergence of heritage studies as a field of inquiry during the 1980s see Rodney Harrison, *Heritage: Critical Approaches* (London: Routledge, 2013), 98.
7 For the integrated conservation approach in international recommendations in the 1970s and 1980s see UNESCO, *Recommendation concerning Safeguarding and*

Contemporary Role of Historic Areas (Paris: UNESCO, 1976), accessed March 3, 2016, http://portal.unesco.org/en/ev.php-URL_ID=13133&URL_DO=DO_TOPIC&URL_SECTION=201.html and ICOMOS, *Charter for the Conservation of Historic Towns and Urban Areas* (Washington: ICOMOS, 1987), accessed March 3, 2016, www.icomos.org/charters/.

8 World Heritage Cities Programme, accessed October 15, 2015, http://whc.unesco.org/en/cities/. In this research ICOMOS is credited with the status of an individual stakeholder, which it rightly possesses, while at the same time, together with the two other advisory bodies, IUCN and ICCROM, and the World Heritage Committee and the World Heritage Centre, it forms an integral part of what in this research is called the "World Heritage organization."

9 For example, Christina Cameron and Mechtild Rössler, *Many Voices, One Vision: The Early Years of the World Heritage Convention* (Farnham, Surrey: Ashgate, 2013), otherwise a thorough account of the implementation of the World Heritage Convention, does not address the process of site rejection. For a brief account by ICOMOS, see Jukka Jokilehto et al., *What is OUV? Defining the Outstanding Universal Value of Cultural World Heritage Properties*. Monuments and sites XVI (Berlin: Hendrik Bäßler Verlag, 2008), 45–46.

10 Jukka Jokilehto et al., *The World Heritage List: Filling the Gaps – an Action Plan for the Future* (München: ICOMOS, 2005), 9.

11 See, for example, Maaria Seppänen, *Global Scale, Local Place? The Making of the Historic Centre of Lima into a World Heritage Site* (Helsinki: University of Helsinki, 1999); Graeme Evans, "Living in a World Heritage City: stakeholders in the dialectic of the universal and particular," *International Journal of Heritage Studies* 8: 2 (2002): 117–136; Carina Green, *Managing Laponia: A World Heritage as Arena for Sami Ethno-Politics in Sweden* (Uppsala: Uppsala University, 2009).

12 Peter Howard, "Historic landscapes and the recent past: whose history?" in *Valuing Historic Environments*, ed. Lisanne Gibson and John Pendlebury (Surrey: Ashgate, 2009), 51.

13 The term "heritage" began to emerge in the discussions at UNESCO during the 1960s when it replaced expressions such as "historical monuments" and "historic and artistic groups and areas." Since its adoption by UNESCO, heritage has become commonly used in a variety of national contexts when explaining the means by which people in the present associate themselves with people in the past. However, as Astrid Swenson notes, the shift from the term "monument" to "heritage" had already started during the last decades of the nineteenth century in connection with the universalization of a moral duty attached to preservation. Astrid Swenson, *The Rise of Heritage: Preserving the Past in France, Germany and England 1789–1914* (Cambridge: Cambridge University Press, 2013), 331–332. See also Graeme Davison, "Heritage: from patrimony to pastiche," in *The Heritage Reader*, ed. Graham Fairclough et al. (London: Routledge, 2008), 31–33, who discusses the adoption of word "heritage" into an Australian context. In the conservation management framework, the concept of heritage is usually divided into cultural and natural heritage, both of which can be associated with tangible and intangible values. Cultural heritage is further divided into the tangible and the intangible. All these divisions have recently been called into question with regard to their strict dualism.

14 Jennifer Jordan, *Structures of Memory: Understanding Urban Change in Berlin and Beyond* (Stanford: Stanford University Press, 2006), 2.

15 For discussion of the nature of heritage see, for example, Eric Hobsbawn, "Introduction: inventing traditions," in *The Invention of Tradition*, ed. Eric Hobsbawn and Terence Ranger (Cambridge: Cambridge University Press, 1983); David Lowenthal, *The Past is a Foreign Country* (Cambridge: Cambridge University Press, 1985); Barbara Kirshenblatt-Gimblett, *Destination Culture. Tourism, Museums, and Heritage* (Berkeley: University

of California Press, 1998); David Lowenthal, *The Heritage Crusade and the Spoils of History* (Cambridge: Cambridge University Press, 1998); Bella Dicks, *Culture on Display: the Production of Contemporary Visitability* (Buckingham: Open University Press, 2003); G. J. Ashworth and B. Graham, "Senses of Place, Senses of Time and Heritage," in *Senses of Place, Senses of Time*, ed. G. J. Ashworth and B. Graham, 3–12 (Aldershot: Ashgate, 2005); Smith, *Uses of Heritage*; Harrison, *Heritage: Critical Approaches*; Sharon Macdonald, *Memorylands. Heritage and Identity in Europe Today* (New York: Routledge, 2013).

16 I use "conservation" to describe a range of interventions in the physical integrity of buildings and places considered to have historical value, such as preservation, restoration, reconstitution, reconstruction, or adaptive use. For the terminology see Harsha Munasinghe, *Urban Conservation and City Life: Case Study of Port City of Galle* (Oulu: University of Oulu, 1998), 12; Rudy Koshar, *Germany's Transient Pasts. Preservation and National Memory in the Twentieth Century* (Chapel Hill: The University of North Carolina Press, 1998), 15. I consider conservation to be active procedures of meaning-making. See also Peter Borsay, *The Image of Georgian Bath, 1700–2000: Towns, Heritage, and History* (Oxford: Oxford University Press, 2000), 374.

17 Smith, *Uses of Heritage*, 3; D. C. Harvey, "A History of Heritage," in *Ashgate Research Companion to Heritage and Identity*, ed. Brian Graham and Perter Howard (Abingdon: Ashgate, 2012), 19–36; Harrison, *Heritage: Critical Approaches*, 14.

18 Concerning land ownership and use as part of memory production see Jordan, *Structures of Memory*, 12–14.

19 Swenson, *The Rise of Heritage*, 2.

20 Colin Long and Sophia Labadi, "Introduction," in *Heritage and Globalisation*, ed. Sophia Labadi and Colin Long (London: Routledge, 2010), 2.

21 Ashworth and Graham, "Senses of Place," 6. The term "dissonant heritage" was introduced by J. E. Tunbridge and G. J. Ashworth, *Dissonant Heritage: the Management of the Past as a Resource in Conflict* (Chichester: Wiley, 1996).

22 See, for example, Lisanne Gibson, "Cultural Landscapes and Identity," in *Valuing Historic Environments*, ed. Lisanne Gibson and John Pendlebury (Surrey: Ashgate, 2009), 74; Labadi, "Representations of the Nation," 153.

23 For a discussion about the key role of determining cultural significance in heritage management see Erica Avarami, Randall Mason and Marta de la Torre, *Values and Heritage Conservation: Research Report* (Los Angeles, CA: Getty Conservation Institute, 2000), accessed February 24, 2016, http://hdl.handle.net/10020/gci_pubs/values_heritage_research_report; Randall Mason, "Assessing Values in Conservation Planning: Methodological Issues and Choices," in *Assessing the Values of Cultural Heritage*, ed. Marta de la Torre (Los Angeles: The Getty Conservation Institute, 2002), 5–30.

24 For discussion within the framework of the World Heritage Convention see Chapter 2 and Labadi, *UNESCO*, 25–58.

25 Mason, "Assessing Values," 8, 13. See also Munasinghe, *Urban Conservation and City Life*; Lisanne Gibson and John Pendlebury, "Introduction: Valuing Historic Environments," in *Valuing Historic Environments*, ed. Lisanne Gibson and John Pendlebury (Surrey: Ashgate, 2009), 7.

26 Laurajane Smith, "Deference and humility: the social values of the country house," in *Valuing Historic Environments*, ed. Lisanne Gibson and John Pendlebury (Surrey: Ashgate, 2009), 35; Lisanne Gibson, "Cultural Landscapes and Identity," 72.

27 Gibson and Pendlebury, "Introduction: Valuing Historic Environments," 10; William S. Logan, "Globalizing Heritage: World Heritage as a Manifestation of Modernism and Challenges from the Periphery," in *20th Century Heritage: Our Recent Cultural Legacy: Proceedings of the Australia International Council for Monuments and Sites National Conference 2001*, ed. David Jones (University of Adelaide & Australia ICOMOS: Adelaide, 2001), 54–55.

28 Labadi, *UNESCO*, 18.
29 Ibid., 18–22.
30 Brian Graham, G. J. Ashworth and J. E. Tunbridge, *A Geography of Heritage. Power, Culture and Economy* (London: Arnold, 2000) deals explicitly with the geography of heritage and the existence and relationships of different levels of heritage. See also Lowenthal, *The Heritage Crusade*, 227–247.
31 See, for example, Hobsbawn, "Introduction: inventing traditions"; Koshar, *Germany's Transient Pasts*; Joshua Hagen, *Preservation, Tourism and Nationalism. The Jewel of the German Past* (Adlershot: Ashgate, 2006).
32 Swenson, *The Rise of Heritage*, 4.
33 See, for example, Koshar, *Germany's Transient Pasts*; Jo Littler, "Introduction: British heritage and the legacies of 'race,'" in *The Politics of Heritage: The Legacies of 'Race,'* ed. Jo Littler and Roshi Naidoo (London: Routledge, 2005); Stuart Hall, "Whose heritage? Un-settling 'the Heritage,' re-imagining the post-nation," in *The Politics of Heritage: The Legacies of 'Race,'* ed. Jo Littler and Roshi Naidoo (London: Routledge, 2005); Smith, *Uses of Heritage*; Emma Waterton, Laurajane Smith and Gary Cambpell, "The Utility of Discourse Analysis to Heritage Studies: The Burra Charter and Social Inclusion," *International Journal of Heritage Studies* 12: 4 (2006): 339–355.
34 Hall, "Whose heritage?," 25; Waterton, Emma, "Branding the past: the visual imagery of England's heritage," in *Culture, Heritage and Representation: Perspectives on Visuality and the Past*, ed. Emma Waterton and Steve Watson (Surrey: Ashgate, 2010), 155.
35 Smith, *Uses of Heritage*, 11; Waterton, Smith and Campbell, "The Utility of Discourse Analysis," 339.
36 Smith, *Uses of Heritage*, 95–102.
37 On discussion concerning the relationship of material and discursive aspects related to heritage see Harrison, *Heritage: Critical Approaches*, 110–111. Smith, *Uses of Heritage*, 74, proposes that the tension between the idea of the intangibility of heritage and the critical reality that determines that there are physical things or 'places' we name and define as heritage is a central aspect of 'heritage.'
38 Concerning the three-layered production of space: spatial practice (the spatial practice of a society presupposes that society's space), representation of space (conceptualized space; space of planners) and representational spaces (space as directly lived) see Henri Levebvre, *The Production of Space*, transl. Donald Nicholson-Smith (Oxford: Blackwell Publishing, 1991), 38–39. For landscapes, see Maunu Häyrynen, *Kuvitettu maa. Suomen kansallisen maisemakuvaston rakentuminen* (Finnish Literature Society: Helsinki, 2005), 27.
39 Margaret C. Rodman, "Empowering place: multilocality and multivocality," in *The Anthropology of Space and Place: Locating Culture*, ed. Setha M. Low and Denise Lawrence-Zúñiga (Malden, MA: Blackwell, 2003), 205.
40 Some sites in the World Heritage List are called 'mixed,' which means that they are considered to have both natural and cultural qualities of outstanding universal value.
41 Leonardo Benevolo, "The city as an expression of culture: the case of 14th century Urbino," in *Partnerships for World Heritage Cities: Culture as a Vector for Sustainable Urban Development*. World Heritage Papers 9 (Paris: World Heritage Centre, 2003), 17.
42 These are the Cultural Landscape of Sintra (Portugal, 1995), the Hallstatt-Dachstein Salzkammergut Cultural Landscape (Austria, 1997), Portovenere, Cinque Terre, and the Islands (Palmaria, Tino and Tinetto) (Italy, 1997), Ferrara, City of the Renaissance and its Po Delta (Italy, 1999), Ibiza, biodiversity and culture (Spain, 1999), and Dresden Elbe Valley (Germany, 2004).
43 See, for example, UNESCO, Bureau of the World Heritage Committee, Twenty-Fourth Session (Paris, 26 June – 1 July 2000), Item 6.3 of the Provisional Agenda: Report of the Working Group on the Representativity of the World Heritage List, WHC-2000/CONF.202/10, 15 May 2000, Annex III: Analysis of the World Heritage List by Category

of Monument and Period (List at January 2000), http://whc.unesco.org/archive/2000/whc-00-conf202-10e.pdf; Jokilehto et al., *The World Heritage List: Filling the Gaps*, 31–48.

44 UNESCO, World Heritage Committee, Twenty-Fourth Session (Cairns, Australia 27 November – 2 December 2000), Evaluations of Cultural Properties. Prepared by the International Council on Monuments and Sites, WHC-2000/CONF.204/INF.6, http://whc.unesco.org/archive/2000/whc-00-conf204-inf6e.pdf.

45 ICOMOS evaluation for the nomination of the World Heritage property, "Spanish Town," Jamaica, No 459 Rev, September 1988. World Heritage Centre, World Heritage Sites, Nomination files (1978–1999), Rejected (CD in author's possession).

46 Tentative Lists, Jamaica, accessed January 29, 2016, http://whc.unesco.org/en/tentativelists/state=jm.

47 UNESCO, Bureau of the World Heritage Committee, Twenty-Second Session, Paris, 22–27 June 1998, Report of the Rapporteur, WHC-98/CONF.201/9 (Paris, 11 August 1998), p. 31, http://whc.unesco.org/archive/1998/whc-98-conf201-9e.pdf.

48 Jokilehto et al., *What is OUV?*, 45.

49 The ICOMOS evaluations concerning accepted urban nominations have been retrieved from http://whc.unesco.org/en/list. Later on in this book they are referred to by the relevant reference number, the name of the site/country and the date of ICOMOS evaluation. The ICOMOS statements concerning rejected urban nominations have been collected from various sources. The ICOMOS Archives do not hold any record of two negative ICOMOS statements issued in 2000, that is, those concerning Santa Fe de Bogotá (Colombia) and Santarém (Portugal). It has thus been impossible to discuss these two cities. In addition, there are a few negative statements which have been consulted from the later positive statement issued by ICOMOS.

50 Jokilehto, *A History*, 296.

51 Waterton, Smith and Campbell, "The Utility of Discourse Analysis." On Critical Discourse Analysis see, for example, Norman Fairclough, *Analysing Discourse: Textual Analysis for Social Research* (London: Routledge, 2003); Norman Fairclough, "Discourse analysis in organizational studies: the case for critical realism," *Organization Studies* 26 (2005): 915–939; Gunther Kress and Theo van Leeuwen, *Reading Images: The Grammar of Visual Design* (London: Routledge, 2006).

52 Waterton, "Branding the past," 156–157.

53 Fairclough, *Analysing Discourse*, 8.

54 Waterton, Smith and Campbell, "The Utility of Discourse Analysis," 346.

55 Ibid., 351.

56 Fairclough, *Analysing Discourse*, 139.

57 Waterton, Smith and Campbell, "The Utility of Discourse Analysis," 344–45. For the different degrees of dialogicality (attribute/quotation; modalized assertion; non-modalized assertion; assumption) see Fairclough, *Analysing Discourse*, 46–47.

58 For oral history on the research of urban history see Rodger, Richard and Herbert, Joanna, eds., *Testimonies of the City: Identity, Community and Change in a Contemporary Urban World* (Aldershot: Ashgate, 2007).

2 Intrinsic value, uncontestable expertise

ICOMOS, UNESCO and outstanding universal value

The nomination of new sites to the World Heritage List is the sole responsibility and right of the States Parties, making them important gatekeepers of global heritage.[1] As many scholars have convincingly shown, the national level nomination agendas are susceptible to several forms of national-political negotiations, and over the years World Heritage has been readily used for purposes of political legitimization and projects of national identity. In addition, various spatial-political and economic factors determine the nomination politics at the national level.[2] A country's inability to nominate sites and to compile a tentative list can be further affected by issues such as limited financial, technical and human resources.[3] It is thus obvious that many concerns other than value assessments related to the establishment of outstanding universal value characterize the building of the World Heritage List. This notwithstanding, outstanding universal value remains the key concept in the World Heritage system.

This chapter explores the concept of outstanding universal value, as outlined by ICOMOS and UNESCO in general level policy statements, as well as the more specific value articulations as displayed in the official World Heritage criteria. As pointed out in Chapter 1, the meaning of outstanding universal value has been (re)constructed over time through several interrelated and case-by-case processes by various stakeholders when putting the Convention into practice. States, in the nomination dossiers, articulate their understanding of outstanding universal value for each proposed site. These documents – today extensive, visually pleasing and often written by consultants – are "the voice of States Parties"[4] in the system. Another important instrument is the tentative lists that states compile of sites that they are planning to nominate in the near future. Whether the voice will be heard or not depends on the World Heritage Committee, the 21 members of which are elected by a General Assembly of States Parties. The Committee has several responsibilities and powers, amongst them the regulation of the composition of the World Heritage List. It decides on whether the nomination should be inscribed on the World Heritage List; rejected; referred back to the State Party for additional information; or deferred for a substantial revision (Table 2.1). While it is for the states to propose and demonstrate outstanding universal value and for the World Heritage Committee to make the final decisions, it is for the two NGOs, ICOMOS and IUCN, to examine if and how this condition is met. Analyzing the intermediary role of ICOMOS and

Table 2.1 The World Heritage site inscription process

---

1 Tentative lists
States compile tentative lists of cultural and natural sites which they consider to be of outstanding universal value and which they are planning to nominate to the World Heritage List during coming years. Tentative lists should be updated regularly, at least every ten years. The present Operational guidelines encourage tentative lists to be prepared in consultation with broader (local, regional, NGO) interests. A site not included in the tentative lists is not considered for inscription by the World Heritage Committee.

2 World Heritage nominations
The State Party decides which tentative listed sites it will nominate to the World Heritage List, and when (if ever). The compiling of the nomination dossier and the related management plan, which follows a pre-structured format, is coordinated by national authorities before submitting them to the UNESCO World Heritage Centre. Nominations are also encouraged to be written in consultation with broader (local, regional, NGO) interests. States Parties may submit draft nominations to the Secretariat/World Heritage Centre for comment.

3 Evaluation of nominations by the advisory bodies (ICOMOS and IUCN)
The Advisory Bodies will evaluate whether or not properties nominated by States Parties have outstanding universal value and meet the conditions of integrity, authenticity and proper management. The evaluations are made on the basis of the criteria defined by the World Heritage Committee. The Advisory Bodies make recommendations to the World Heritage Committee.

4 World Heritage Committee decision-making
The World Heritage Committee decides whether a nominated site is or is not inscribed on the World Heritage list. Two other decision categories are a referral back to the State Party and a deferral.

---

its evaluations in the World Heritage system is another key concern of this chapter. The chapter begins with a discussion placing the World Heritage Convention in the context of the postwar globalizing and modernizing heritage concern.

The World Heritage Convention in context

The birth of the World Heritage idea was part of the international community's response to the growing tendencies of modernization, globalization and "heritagization" during the 1960s.[5] The signing of the World Heritage Convention was a result of a long period of earlier international cooperation, as illustrated by Sarah Tichen,[6] Christina Cameron and Mechtild Rössler[7] in their thorough step-by-step accounts of the preparation phase. In its concern for the common heritage of humankind, post-Second World War UNESCO had a direct predecessor in the post-First World War League of Nations, which, in the interwar years, had promoted the concept of a common cultural heritage as part of its effort to advance international interests over nationalistic ones.[8] The concept of a common heritage, however, dates further back: Hugo Grotius (1583–1645), a statesman, jurist and scholar of Dutch origin, applied this notion to the world's seas. During the eighteenth and nineteenth centuries increasing attention was paid to the fate of art works in times of war, with those works understood as being part of human-kind's heritage.[9] What is often forgotten is that despite strong nationalist underpinnings, by the late nineteenth century heritage had already become an international concern.[10]

A resolution adopted by the general assembly of the International Alliance of Tourism in 1949 requested international organizations to ensure the safe-guarding of monuments, "which are the common heritage of all civilized nations, by bringing their traditional laws in line and by organizing as a matter of urgency international financial assistance."[11] This reference shows that the idea of humankind's common heritage had by then found broader resonance, and was not restricted to a small circle of heritage professionals. Since the late 1940s the ideas and practices concerning the worldwide protection of monuments and natural environments, at the time still viewed separately, evolved at an accelerating speed. UNESCO's initiatives included the ratification of the Hague Convention in 1954, the founding of the International Centre for the Study of the Preservation and Restoration of Cultural Property (ICCROM) in 1959, and the international campaigns to save the Nubian monuments in Egypt and Sudan and the flooding Italian cities of Florence and Venice, launched in 1960 and 1966 respectively. In 1964 the milestone Second Congress of Architects and Technicians of Historic Monuments met in Venice. This meeting produced three results, all of which played an important role in the establishment of an internationally based heritage profession and discourse with the intent of guiding practices at the national level. First, it adopted the *International Charter for the Conservation and Restoration of Monuments and Sites*, commonly known as the Venice Charter. Second, with the support of UNESCO, the meeting initiated the founding of ICOMOS a year later in Warsaw. Third, the Congress led to the establishment, within ICCROM, of an international training program for conservation.[12]

All these institutions – ICOMOS, ICCROM and the Venice Charter – may be considered highly significant for the later operation of the World Heritage Convention. Both ICOMOS and ICCROM were given strong roles as advisory organizations of UNESCO in all matters dealing with cultural World Heritage: ICOMOS in questions concerning heritage valuation and management, and ICCROM in training issues. Both these organizations, and the authors of the Venice Charter, also participated in the writing of the World Heritage Convention. By the time of the Convention's drafting, the Venice Charter had already been canonized as "in all the world the official code in the field of the conservation of cultural properties."[13] Consequently, it became the main underlying reference document for value assessment applied within the World Heritage framework. Whilst still referred to among heritage professionals, the Venice Charter has evoked a fair amount of criticism. This has primarily been aimed at the Charter's monumental view regarding heritage, its overemphasis on Western concepts of material permanency and authenticity in conservation, its unequivocal condemnation of "pseudo-historical design," and its underlying modernist rhetoric concerning universalizing concepts and a desire for classification.[14] As might be anticipated, it is possible to voice very similar critiques concerning the World Heritage Convention, since the two documents are so closely bound together.

A great deal of the inner logic of the World Heritage idea has to do with protecting local diversities and anti-modern originals from the expected harmful and culturally unifying effects of globalization and modernization. Risk is the fundamental notion to the experience of modernity, and thus to modern conservation;[15] similarly World Heritage was essentially defined in the context of threat to places, as made explicit in the Convention text.[16] Also, the dual system of listing of World Heritage sites is based on the same notion of heritage at risk: the broad World Heritage List is supplemented by the short List of World Heritage in Danger. In the professional language of international heritage specialists the 'bad' homogenizing effects of globalization have often been separated from the desirable forms of institutionally guided globalization. Obviously, the World Heritage system itself is also part of the same process of homogenizing globalization that it aims to counteract.[17] Stating that a local place and a local culture is a common World Heritage subject creates, in itself, a globalizing effect, as do the standardized processes of nomination, evaluation and designation.

The attempts to preserve the original through World Heritage designation may lead, paradoxically, to hopes for and efforts toward modernization in the local context.[18] Furthermore, as with any heritage protection, World Heritage, while essentially stemming from a critique of modernization in society, has largely operated within the very same modernist discourse and modernist logic, based, for example, on an objectified understanding of the human past, reliance on a hierarchical listing exercise, heritage categorizations, scientifically set criteria and the role of experts, linked to the idea of ever continuous growth (in size, amount, density, etc.) of protected sites. As noted by Lisanne Gibson and John Pendlebury, the established processes of heritage "are usually inherently 'modern' for two reasons. First, they are a reaction to the threat caused by progressive modernity and the change (whether aesthetic or social) that this implies. [...] Second, heritage professionals are people of the modern age. Their concepts of history and cultural value and their methods of pursuing their goals are as intrinsically modern as those of the promoters of change."[19]

The World Heritage Convention is a wide-ranging document, containing a preamble and 38 paragraphs. In these paragraphs, the Convention articulates its understanding of heritage, explains how World Heritage should be protected, outlines conditions for international assistance and establishes the different institutions for the implementation of the Convention, most importantly the World Heritage Committee and the World Heritage Fund. Furthermore, it discusses, somewhat ambiguously, the relationship between the concept of world heritage and the sovereign rights of the signatory states.[20] The definition of heritage introduced in the Convention text was progressive in the context of the early 1970s. Not only did it treat heritage as both cultural and natural heritage, thus recognizing the inevitable link between the two, but it also introduced an advanced definition of cultural heritage. The term used in the Venice Charter was "historic monument,"[21] and in the mid-1960s, UNESCO also spoke about "monuments of historical and artistic value" as subjects in need of protection.[22] A few years later, the drafting group of what was to become the World Heritage Convention, introduced a more extensive terminology,[23] which in Article 1 of the World Heritage Convention was combined under the three categories of "monuments," "groups of buildings" and "sites." Groups of buildings were understood to comprise "groups of separate or connected buildings which, because of their architecture, their homogeneity or their place in the landscape, are of outstanding universal value from the point of view of history, art or science" and the sites as "the works of man or the combined works of nature and man."[24] In addition to this fairly broad conception of cultural heritage, the Convention spoke primarily, and contrary to other UNESCO conventions and recommendations at the time, in terms of heritage, and not of property, something which may be seen as an attempt to focus on a global sense of duty and stewardship in opposition to national ownership.[25]

The universal claim of value

In addition to the case-by-case construction of outstanding universal value, there have been continuing efforts on the part of the World Heritage Committee and its advisory organizations to provide a more general explanation of the concept. An example of this effort is the comprehensive report on the historical development and current policy definitions related to the concept drawn up by ICOMOS in 2008.[26] Here I will not attempt to reproduce the discussions concerning outstanding universal value in the World Heritage framework; my aim, instead, will be to summarize the main trends visible in these conversations over time, and by doing so explore both the permanent and evolving elements of the concept. To begin with, it should be pointed out that there has been, for the entire time that the Convention has been in existence, a wide consensus about the position of outstanding universal value as a fundamental condition for the definition of World Heritage, despite the fact that some engaged in the World Heritage practice, such as the former ICOMOS World Heritage coordinator, Henri Cleere, have denounced the concept as "too vague" to serve as a defining condition for World Heritage listing.[27] The doubts, such as those expressed by Cleere, have especially been raised concerning the word universal, if understood in terms of universally shared values, which in the framework of post-modern pluralization of value are difficult to justify. A place, even though valued by many, can be valued for different reasons and based on different values.[28] Outstanding universal value has also been criticized for being a Western construction, a concept, which "reinforces Western notions of value and rights."[29]

The universalist view has not been uncommon in the course of the implementation of the Convention. For instance, the group of experts that convened in 1976 to draft the first cultural criteria noted that "a property submitted for inclusion in the World Heritage List should represent or symbolize a set of ideas or values which are universally recognized as important, or as having influenced the evolution of mankind as a whole at one time or another."[30] At the same time, it should be noted that even at that early stage, more culture-specific conceptions of outstanding universal value were also defined, as was the case in the first session of the World Heritage Committee in 1977, where it was stated that "the term 'universal' must be interpreted as referring to a (cultural) property which is highly representative of the culture of which it forms part."[31]

Those advancing the concept of outstanding universal value in the World Heritage framework more recently have resolved the universalist dilemma by providing an interpretation of the term "universal" as something common to all human cultures, themselves diverse.[32] In other words, outstanding universal value should be understood as "an outstanding response to issues of universal nature common to or addressed by all human cultures," and to be found "in human creativity and resulting cultural diversity."[33] There continue to be those sites that are "most outstanding," but the reflection should be carried out within the cultural context, not on a universal scale.[34] Still, even in recent years those engaging themselves in the definition of outstanding universal value have not been entirely unanimous in their articulation of the concept. For example, the President of ICOMOS International between 1999 and 2008, Michael Petzet, stated in 2008 that universal "means that these outstanding values can be acknowledged as such in general and worldwide."[35] It is possible to find a similar understanding of outstanding universal value in Christina Cameron's keynote speech that she gave in the expert meeting on the concept of outstanding universal value in Kazan in 2005. Cameron, the chairperson of the World Heritage Committee in 2007–2008, asserted that the discussions involving outstanding universal value have focused on two interpretations of the concept: "the best of the best" and "representative of the best." According to Cameron, there has been a shift from the early practice of World Heritage listing of iconic sites clearly meeting "the benchmark of 'best of the best'" towards the listing of "representative of the best."[36] It is true that while adopting the Global Strategy based on a wide thematic comparative approach (see Chapter 3), the World Heritage Committee chose the policy of selecting representative sites. However, the understanding that a good number of iconic, "best of the best" sites were added to the World Heritage List in the early years of the Convention suggests that it was then possible to identify sites having intrinsic universal value. However, when looking at the early stage ICOMOS evaluation texts, it becomes obvious that there were many nominations which were not even at that time considered to represent "the best of the best." Moreover, the standards for "the best" were measured and met from a Western perspective.[37]

The broadly shared understanding of outstanding universal value among World Heritage practitioners, has, over time, been that the concept should operate as a control mechanism for maintaining the credibility of the World Heritage List. The credibility, for its part, has been understood to mean two things: on the one hand, broad inclusiveness of different regions, cultures, and types of heritage, i.e. the representativeness of the World Heritage List, and, on the other hand, the selective and distinctive nature of the World Heritage List, not open to all sites.[38] These two aspects could be seen as contradictory to each other, since a high level of selectivity might restrict inclusiveness and vice versa. The coupling of the demand of outstanding universal value with the idea of balance has thus been criticized for inherent incompatibility.[39] Usually in the World Heritage rhetoric this has not been considered a problem, since most often the concern for representativeness of the World Heritage List, and for maintaining a high threshold for outstanding universal value in the process have been expressed simultaneously.[40]

It should also be noted that in the overall World Heritage discourse the outstanding universal value has often been considered as an intrinsic quality of a place and as something that can be objectively determined if only strict enough criteria are established and followed through during a process of scientific evaluation. Similarly, there has generally been a strong conviction about the inherent objectivity and the technical nature of the World Heritage site selection process.[41] In recent years the World Heritage organization has more often acknowledged that outstanding universal value is attributed by people,[42] and consequently, that it is a social construction of time and place. Despite this realization, it may be argued that the idea of outstanding universal value as an intrinsic value has not disappeared from the discourses on World Heritage.[43] The Operational Guidelines for the Implementation of the World Heritage Convention consider outstanding universal value to mean "common importance for present and future generations of all humanity," and continue occasionally to refer to the term "intrinsic."[44] Also, the above-cited expert meeting in Kazan, while acknowledging that outstanding universal value is attributed by people, held to the concept of intrinsic value; the meeting was in fact called together in response to the concern by the World Heritage Committee "that this concept is interpreted and applied differently in different regions and by different stakeholders."[45] These ambiguities in the definition of outstanding universal value are in line with the notion of Lisanne Gibson and John Pendlebury that the official discourses regarding heritage, despite the increasing awareness of heritage values as a social, spatial and historical construction, have clung to ideas of intrinsic values, and to "a wish to separate them from the most obviously instrumental performative roles of heritage."[46] In addition, the methods and the manner in which the outstanding universal value is identified for an individual site have remained largely unaltered over the years. It has been considered that the outstanding universal value may be distinguished by careful definition and rigorous application of criteria, by comparison with already listed and other sites, by defining different types and categories of heritage and by building up global frameworks and databases.[47] In this effort, comparative assessment has been judged to be central.[48] It is within and in relation to the broader comparative frameworks that the outstanding universal value of an individual site should be considered. I deliberately choose the wording "should be," for when looking at individual nominations and evaluations, even the more recent ones, there is a significant divergence in how the comparative assessment has been carried out, especially with regard to the chosen scope of the theme for comparison. It is also important to note that the comparative approach has fit somewhat uneasily with highlighting the site's uniqueness, another important attribute in World Heritage. Thus, the fundamental threshold for measuring outstanding universal value in the World Heritage system is made up of the criteria outlined in the Operational Guidelines.[49] The development of these criteria, and their value basis, will be focused on next.

World Heritage values in perspective

Cultural heritage criteria

Heritage valuation, by experts, is commonly performed in reference to typologies of value, even though concern has also been expressed in regard to these typologies representing a reductionist approach to the examination of cultural significance, which forms in turn a diverse and complex issue.[50] A primary value typology for cultural World Heritage is formulated in the Convention text in association with the definition of the various types of cultural heritage. The main values articulated for monuments and groups of buildings are historical, artistic/aesthetic and scientific. For the category of sites, the Convention text further mentions ethnological and anthropological values.[51] The ethnological and anthropological values, however, have often been understood as scientific values,[52] despite their wider potential. Another value typology for World Heritage is based on the criteria outlined in the Operational Guidelines. Over the years there has been strong confidence in the ability of the World Heritage Committee and ICOMOS to establish firm and universally applicable criteria for the value assessment, as noted above, as well as an awareness of the difficulty of applying them. The members of the Committee in their very first meeting expressed several potential problems with such an undertaking in regard to the inherent subjectivity related to evaluating qualities, the impact of Western thought and the difference between viewing the criteria from within a given culture and from outside, all issues later identified as challenges and even faults of the Convention. The representative of ICOMOS in the meeting noted that an attempt had been made to translate concepts "into words that were meaningful on a universal scale" but that this was a challenging task and that "the criteria would probably require some adjustment" in the future.[53] The organization, nevertheless, decided to tackle this project.

Unlike the Convention text which has remained unaltered since its drafting in 1972, the Operational Guidelines is a document that is regularly modified. This has also resulted in regular revision of the criteria used in the assessment of outstanding universal value. In retrospect, the two most significant periods of modification of the cultural criteria occurred around the years 1980 and 1995 (Table 2.2).

The first criteria of 1977 were relatively wide in their scope. This was especially the case with criteria ii and iv, which mentioned "human settlements" and "social development." These references to social values were removed in the 1980 revision, criterion ii now focusing on developments in architecture and town-planning, and criterion iv referring only to a "type of structure." While these revisions, together with the addition of a reference to a "civilization which has disappeared" in criterion iii, meant restriction in perspective,[54] criterion i in turn was widened in scope by deleting the notion of aesthetic value. It is nevertheless difficult to maintain, as argued by Christina Cameron and Mechtild Rössler, that there was a shift in the definition of cultural criteria "from an architectural or geographic perspective to an anthropological point of view" between 1976 and 1980;[55] the shift occurred rather into a reversed direction. Furthermore, in the

Table 2.2 Comparison of cultural criteria in the assessment of outstanding universal value in 1977, 1980, 1995 and 2005

1977	1980	1997	2005
(i) represent a unique artistic or aesthetic achievement, a masterpiece of the creative genius	(i) represent a unique artistic achievement, a masterpiece of the creative genius	(i) represent a masterpiece of human creative genius	(i) represent a masterpiece of human creative genius
(ii) have exerted considerable influence, over a span of time or within a cultural area of the world, on subsequent developments in architecture, monumental sculpture, garden and landscape design, related arts, or human settlements	(ii) have exerted great influence, over a span of time or within a cultural area of the world, on developments in architecture, monumental arts or town-planning and landscaping	(ii) exhibit an important interchange of human values, over a span of time or within a cultural area of the world, on developments in architecture or technology, monumental arts or town-planning and landscape design	(ii) exhibit an important interchange of human values, over a span of time or within a cultural area of the world, on developments in architecture or technology, monumental arts or town-planning and landscape design
(iii) be unique, extremely rare, or of great antiquity	(iii) bear a unique or at least exceptional testimony to a civilization which has disappeared	(iii) bear a unique or at least exceptional testimony to a cultural tradition or to a civilization which is living or which has disappeared	(iii) bear a unique or at least exceptional testimony to a cultural tradition or to a civilization which is living or which has disappeared
(iv) be among the most characteristic examples of a type of structure, the type representing important cultural, social, artistic, scientific, technological or industrial development	(iv) be an outstanding example of a type of structure which illustrates a significant stage in history	(iv) be an outstanding example of a type of building or architectural or technological ensemble or landscape which illustrates (a) significant stage(s) in human history	(iv) be an outstanding example of a type of building or architectural or technological ensemble or landscape which illustrates (a) significant stage(s) in human history

1977	1980	1997	2005
(v) be a characteristic example of a significant, traditional style of architecture, method of construction, or human settlement, that is fragile by nature or has become vulnerable under the impact of irreversible socio-cultural or economic change	(v) be an outstanding example of a traditional human settlement or land-use which is representative of a culture, especially when it has become vulnerable under the impact of irreversible change	(v) be an outstanding example of a traditional human settlement or land-use which is representative of a culture (or cultures), especially when it has become vulnerable under the impact of irreversible change	(v) be an outstanding example of a traditional human settlement, land-use, or sea-use which is representative of a culture (or cultures), or human interaction with the environment especially when it has become vulnerable under the impact of irreversible change
(vi) be most importantly associated with ideas or beliefs, with events or with persons, of outstanding historical importance or significance	(vi) be directly or tangibly associated with events or with ideas or beliefs of outstanding universal significance. (The Committee considers that this criterion should justify inclusion in the List only in exceptional circumstances or in conjunction with other criteria)	(vi) be directly or tangibly associated with events or living traditions, with ideas, or with beliefs, with artistic and literary works of outstanding universal significance. (The Committee considers that this criterion should justify inclusion in the List only in exceptional circumstances and in conjunction with other criteria cultural or natural)	(vi) be directly or tangibly associated with events or living traditions, with ideas, or with beliefs, with artistic and literary works of outstanding universal significance. (The Committee considers that this criterion should preferably be used in conjunction with other criteria)

Sources: Operational Guidelines, 20 October 1977; Operational Guidelines, October 1980; Operational Guidelines, February 1997; Operational Guidelines, 2 February 2005, retrieved from http://whc.unesco.org/en/guidelines/.

1979 Committee meeting, criterion vi had come under particular scrutiny, for its wording was considered too broad in the sense that its extensive use could potentially result in "a reduction in the value of the List, due to the large potential number of nominations as well as to political difficulties."[56] The conclusion was that the main emphasis of the World Heritage List should be on "concrete properties," whose historical importance depended on "tangible features of self-evident quality." What should be avoided was "letting the List become a sort of competitive Honours Board for famous men of different countries."[57] As a response, the reference to associations with persons of outstanding historical importance was deleted from the wording of criterion vi. Also the use of the criterion was significantly restricted, being from this moment onwards considered applicable only "in exceptional circumstances or in conjunction with other criteria." The removal of the reference to persons from the 1980 criteria contributed to the absence of social values in the criteria;[58] the main incentives for this change, however, were, first, to avoid the World Heritage List becoming politicized and, second, to keep the List's focus on tangible aspects of heritage.

The mid-1990s meant a notable reformulation of the cultural criteria in the context of a wider conceptual reassessment (see Chapter 3). Consequently, less importance was placed on uniqueness value (criterion i); there was a transition from one-sided influence exerted by one civilization or dominant culture on another culture to the broader idea of a two-sided interchange of values (criterion ii); sequential stages of development and cultures were improved instead of one dominant historical period and culture (criteria iv and v); and living traditions were acknowledged as part of World Heritage value (criteria iii). However, not all changes in the criteria in the 1990s worked towards broader inclusiveness. Another minor but influential restriction to criterion vi was included in the 1997 Operational Guidelines, as the use of the criterion was limited to only "exceptional circumstances *and* in conjunction with other criteria" rather than the previously stated "or." This re-wording was contrary to the general objective at the time of including less traditional sites in the World Heritage List.[59] In 2005 the original two separate sets of criteria, one for cultural and one for natural heritage, were combined under a joint heading. At this point the individual cultural criteria were altered relatively little. Two significant changes, however, may be noted. The first change involved criterion v, which re-introduced a reference to "human interaction with the environment." The second change again concerned criterion vi, the scope of which was widened after almost a decade of strict interpretation. Since then it has been possible to use this criterion "preferably in conjunction with other criteria." From 2005 until the present the cultural criteria have remained unaltered.[60]

What can be concluded about the value typology laid out in the two basic texts of World Heritage, the Convention and the Operational Guidelines? To be able to answer that question I will now discuss heritage value typologies presented in other contexts. While generally sharing the multiple values perspective, even the recent heritage value typologies differ in their emphases on various values. The discussion draws on the work of several authors, but is primarily indebted to the typology of heritage values presented by Randall Mason.[61]

Assessing cultural heritage significance

Alois Riegl, writing in 1903, already distinguished between multiple heritage values. Riegl's separation of memorial values from present-day values[62] has served as the basis for several later typologies. For example, Bernard M. Feilden and Jukka Jokilehto in their *Management Guidelines for World Cultural Heritage Sites* create a division between cultural values and contemporary socio-economic values. In their scheme, cultural values are further distinguished by identity, rarity and relative artistic and technical values, whereas contemporary socio-economic (use) values include economic, functional, educational, social and political values.[63] In addition to those mentioned by Feilden and Jokilehto, contemporary heritage values could include, for instance, recreational and inspirational values.[64]

Randall Mason distinguishes two major categories of heritage values: the socio-cultural and the economic (Table 2.3). In other words, he does not specify any contemporary values, mainly because he sees all heritage values as contemporary. He also places social values in tandem with cultural values. The difference between the two value sets, socio-cultural and economic, according to Mason, relates to the fundamentally different conceptual frameworks used in their articulation, for economic values are not commensurable with the narrative epistemologies used in the definition of socio-cultural values. This notwithstanding, socio-cultural and economic values may be seen as two overlapping ways of understanding the same wide range of heritage values, and not as entirely discrete sets of values.[65] I will not, however, discuss here economic values, since they have been located completely outside the matrixes used in the World Heritage valuation of cities, except perhaps for a few passing references to tourism. That economic value has been located outside the World Heritage valuation matrix does not mean that economic interests regarding World Heritage sites have not been acknowledged by the World Heritage organization over the years. Often, however, this acknowledgement has emphasized the potential conflict of economic values with other heritage values.[66]

Historical and aesthetic values have a long history of possessing a central position in the definition of cultural heritage significance, also underpinning many other values. It is difficult to think of a heritage place without historical value. As asserted by Mason, "the capacity of a site to convey, embody, or stimulate a relation or reaction

Table 2.3 Randall Mason's provisional typology of heritage values

Socio-cultural Values	Economic Values
Historical	Use market value
Cultural/Symbolic	Nonuse (nonmarket) value
Social	Existence
Spiritual/Religious	Option
Aesthetic	Bequest

Source: Mason, "Assessing Values," 10.

to the past is part of the fundamental nature and meaning of heritage objects."[67] Still, even historical value is not intrinsic to the old material remains from the past, but, instead, something that needs to be negotiated in a given situation.[68] This may be seen, for example, in the way the definition of age value varies in different national contexts. In addition to age, other associations used to define historical value are those connected with important historic events, people and ideas. Artistic and architectural values are also closely related to historical value, even though they are often considered as a separate set of values. Other criteria for measuring historical value include the rarity and documentary potential of a place, criteria which are often also considered to make up a part of scientific values.[69] Moreover, as a recent value typology proposed by English Heritage states, "historical value relates to the ways in which the present can be connected through a place to past people, events and aspects of life,"[70] thus bringing historical value closer to social and associated values.

Beauty in the commonly used sense of the word is not a fundamental requirement for a place to become valued as heritage, because something considered ugly by the majority can still be treated as heritage; however, aesthetic stimulus has always had great power to create interest in an object. While granting that "some forms, textures, and qualities of cultural material are more intrinsically appealing to the observer's aesthetic sense than are others," Walter Lipe summons the complex context from which the aesthetic value of a cultural resource emerges in interplay with the conditions specific to the observer. According to him, the context is "influenced by traditional standards of style and beauty; by critical writings stemming from art history research; by conceptions of what aesthetic standards were held by the culture that produced the item; and by standards deriving from the existence of a market for the type of cultural resource in question."[71] Until recently, what can be considered aesthetic in the context of cultural heritage has primarily been understood in relation to visual perception. However, some recent typologies have widened the aesthetic concept to include other senses as well.[72]

All the other sub-categories that Mason presents under the heading 'socio-cultural values' – cultural/symbolic, social and spiritual/religious – can be considered late-comers to the traditional heritage valuation process. Spiritual/religious value, also encompassing secular experiences of wonder, constitutes a relatively straightforward category by definition, even though not necessarily as part of the valuation process.[73] Cultural/symbolic values, by Mason's definition, are used to "build cultural affiliation in the present" and "can be historical, political, ethnic, or related to other means of living together."[74] Symbolic values can be positive and negative, and perhaps more than any other heritage values, are subject to contestation.

The concept of social value is included in most recent value typologies. According to Chris Johnston, places with social value are those which can join the past and the present; which help to give a disempowered group back its history; which provide an essential community function and a reference point in a community's identity; and which shape some aspect of community behavior or attitudes.[75] Randall Mason regards social values of heritage as something enabling social connections, networks and other relations. As an example of social value he cites the use of a place for such social gatherings as celebrations, markets, picnics or ball games, which associate with "public-space, shared-space qualities." Like Johnston, Mason relates social value with place attachment and community identity, the community ranging from very small and local to national in scale.[76] In these conceptualizations the heritage value of places essentially belongs within the communities, instead of in the visible and tangible fabric. The heritage profession has a long tradition and much expertise in determining historical and aesthetic qualities of places but less experience in identifying social values.[77] More important, even though today they are acknowledged as values in their own right, the question still remains as to whether social values are truly acted on in the processes of heritage assessment and management.

The restricted scope

How do the heritage values articulated as part of the World Heritage display themselves, when reflected against the matrix of heritage values discussed above? To put it briefly, they show a restricted scope, mostly reproducing the traditional value arguments based on historical, artistic and scientific values. This restrictedness particularly involves the definition included in the World Heritage Convention; this is hardly surprising considering the fact that the Convention text has not been redrafted since 1972. What is perhaps more surprising is that the definition given in the Convention has been considered valid by some of the key practitioners of World Heritage, and, more strikingly so, because of the claimed universal applicability of these "classical" historical and aesthetic values.[78] Without denying the centrality of historical and aesthetic/artistic values, it may still be questioned as to why these should be the only values spoken aloud. Another international charter of wide reference, The Burra Charter (1979, 1999), generally acknowledging the relativity of heritage values, defines *cultural significance* in addition to aesthetic, historic and scientific value as social and spiritual value.[79] Also the *Council of Europe Framework Convention on the Value of Cultural Heritage for Society* (2005) begins with a notion about the "need to put people and human values at the centre of an enlarged and cross-disciplinary concept of cultural heritage."[80]

One obvious reason for not being motivated to change the definition given in the Convention text is that opening it to re-formulation would result in a major re-ratification process by the States Parties. This could potentially erode the implementation of the Convention, if some present member states decided not to adhere to the Convention anew. This again places high expectations on the criteria presented in the Operational Guidelines. However, they, too, seem unsuccessful in terms of the articulation of a broad range of heritage values. While it is obvious that the criteria have become more inclusive in terms of manifestations of different cultures of the world, they nevertheless remain imprecise in reference to types of heritage values. The main problem with the criteria is that they actually do not elaborate as much on different heritage values as on attributes to values, for example "a unique or at least exceptional testimony" or "an outstanding example."[81] Whereas some of the terminological ambiguity in the criteria's definitions may be warranted on the basis of an attempt towards inclusiveness, this should not be the case with the articulation of values.

Heritage values are nevertheless implicitly written into the criteria, which for the most part reproduce the historical, artistic/aesthetic and scientific typology outlined in the Convention text. Historical and scientific values may be associated with all of the six criteria, and aesthetic value with at least three of them, criteria i, ii and iv, even though the most direct references to beauty have been removed from the criteria over the years. References to "human" and "living" re-established social value in the mid-1990s; these references, however, remained sparse and undefined. Certain wordings that are used in the criteria have powerful underlying assumptions written into them, in particular the "masterpiece of human creative genius" of criterion i. Furthermore, while criterion vi articulates intangible value, it has been made conditional on other, tangible, values.

As noted above, the *Management Guidelines for World Cultural Heritage Sites* identify a broad spectrum of heritage values relevant to the management of World Heritage sites under the headings of cultural and contemporary socio-economic (use) values.[82] From the manner in which those different values are discussed in the Guidelines, it becomes apparent that whereas outstanding universal value is built on relative artistic, historical and technical values, as defined in an expert valuation, the other values, contemporary socio-economic, are treated mostly as problematic and as something that should be managed, especially when they conflict with historical and aesthetic values. When looking at the articulation of heritage values in the Convention text and through the cultural criteria, one is bound to draw a similar conclusion. In this light, the evolvement of cultural criteria over the years does not show itself to be something revolutionary.

At the same time the enlarged scope of the criteria should be acknowledged. Reformulating the criteria on a regular basis has enabled the World Heritage Committee to recognize the evolving concepts of heritage in society, even though there is often a time lag in comparison with the most innovative systems in certain member countries. The current wordings of criteria iii and v, with their notions of "traditional human settlement," "human interaction with the environment" and "a cultural tradition which is living," express the relevance of cultural and social values. The widening scope is especially notable with regard to criterion iii – while before 1995 the wording was "to bear a unique or at least exceptional testimony to a civilization which has disappeared," it today refers to living cultural traditions. The important issue is, on the one hand, how the possibilities of the different criteria have been utilized, and how, on the other hand, their restrictions have been dealt with in day-to-day valuation practice. These questions will be addressed in Chapters 4 and 5.

Constructing outstanding universal value: the expertise of ICOMOS

Unquestioned objectivity

ICOMOS is a non-governmental organization whose membership comprises professionals and academics from all over the world. It is in this international scope that the political legitimacy and lobbying power of ICOMOS lies.[83] Within the

World Heritage framework, the professional evaluation of nominations has been the focus of agendas of ICOMOS – and IUCN – from the very beginning, as has been the (re)drafting of the criteria used in this assessment, together with the World Heritage Committee. Their tasks soon expanded to other related fields of expertise, also affecting the definition of outstanding universal value: from compiling typologies and comparative studies of different kinds of heritage and assisting State Parties in making tentative lists and nominations, to monitoring and reporting on the state of conservation of already listed sites. Both also provide the Committee with analyses and reports on philosophical, methodological and practical aspects of the Convention.[84]

In the late 1970s everything that involved World Heritage was new to everyone, including ICOMOS. In 1978 all evaluations of that year's nominated cultural heritage sites were covered in a single letter by Ernst Connally, Secretary General of ICOMOS, with technical forms concerning each individual site (titled "ICOMOS technical review notes").[85] The evaluations were presented orally to the World Heritage Committee. The same practice continued the following year; curiously enough the official files for the 1979 inscribed sites were produced only later.[86] According to the director of the ICOMOS International Secretariat in 1979–1983, François LeBlanc, the World Heritage Committee meeting in Luxor, Egypt, in 1979, marked a turning point in the participation of ICOMOS in the World Heritage Convention. At that meeting the Committee members asked ICOMOS to provide more detailed justifications for the reasons for its recommendations for inclusion or exclusion of cultural properties in the World Heritage List. The Committee also called for the organization to be stricter in the future when applying the criteria of outstanding universal value. In the same meeting ICOMOS was also granted some additional financial means to accomplish its task. The representatives of ICOMOS came back to Paris from the Committee meeting, as Leblanc puts it, "with a clear objective and the means to attain it."[87] It was decided to appoint a World Heritage coordinator within ICOMOS to examine cultural nominations. It also became the duty of the coordinator to "centralize" the scholarly opinion obtained in consultation with relevant experts before presenting his conclusions to the seven-member ICOMOS Bureau, which would then decide on the final ICOMOS recommendation for each nomination to be forwarded to the World Heritage Committee.[88]

The writing of the early assessments of outstanding universal value was thus confined to a small group of ICOMOS scholars. This policy resulted in part from the limited resources at ICOMOS' disposal and the multiple priorities of the organization,[89] but it was also believed to increase the objectivity and credibility of the evaluation process. The role of the first World Heritage coordinator, Léon Pressouyre, a French archaeologist and art historian, acting as the coordinator between 1980 and 1990, was very prominent with regard to the cultural evaluations during the 1980s. To make his assessment of a nomination file produced by a State Party, Pressouyre sometimes consulted other experts in the ICOMOS network, but in certain other cases, either with famous sites such as Rome or with sites he personally knew well, no such consultation was considered necessary. Furthermore, the sites under evaluation were actually visited only in rare instances. At that time ICOMOS was of the opinion that even its own specialized Scientific Committees or its National Commissions should not be involved in the evaluation process in any formal way. It was thought, in the case of the former, that these experts might be partial and risk the credibility of the World Heritage List in favor of their specialized fields, and in the case of the latter, that their views could be undermined by the political power of the States Parties.[90] There was strong faith in the objectivity of the evaluation process if just a minimal number of experts were involved.

A renewed request to the advisory bodies for higher standards and objectivity, in order to avoid "favouritism or prejudice," was made in 1983, by professor Ralph Slatyer, the ongoing chairman of the World Heritage Committee.[91] When speaking at the Bureau of the World Heritage Committee meeting earlier the same year, the President of ICOMOS from 1981–1987, Michel Parent, raised the issue of the difficulty in drawing the line between accepted and rejected nominations: "It is a fact that the ever more numerous nominations give rise to contradictory evaluations, and we find ourselves involved at present with a great number of nominations where those 'above the threshold' are so close to those 'below the threshold' that the opinion adopted by a majority of a given body on one occasion could well be different on another occasion, with only a slight change in the body's composition."[92] Yet the main message of Parent's speech was something other than making visible the multiple, and sometimes opposing, views involved in the World Heritage evaluation procedures. The purpose was to argue in favor of developing more rigorous methods in order to attain better (complete) objectivity in defining outstanding universal value in the future. The best way to accomplish this goal was to compile comparative studies based on regional, thematic or even global reflection. In the words of Parent, without them the Committee decisions "would have neither objectivity nor scientific or moral foundation."[93]

Towards broader expert judgment

By the early 1990s, the challenges and inconsistencies of ICOMOS' task became subject to growing concern both among the States Parties and within ICOMOS' own executive committee. Several States Parties voiced critical concerns concerning Léon Pressouyre's dominant role in preparing the evaluations. Pressouyre decided to resign, leading ICOMOS into a fundamental debate about its participation in World Heritage work.[94] An initiative was born to re-constitute the evaluation procedures to include a broader expertise and to sharpen the selection of World Heritage sites. In 1991 ICOMOS introduced to the World Heritage Committee several proposals on how it wished to review future nominations. These included, for example, the suggestion that all sites should be visited by specialists before the debate, and that the National Commissions and the International Scientific Committees of ICOMOS should be associated with the evaluation process.[95] This meant an evolvement of the earlier policy of very limited expert involvement towards a broader expert judgment. It took some time before the new principles introduced in 1991 became standard practice. For example, in 1995, even though most of the historical cities then being evaluated were visited by an ICOMOS mission, the Turkish town of Safranbolu was not. Instead it was reported to be well known to several members of the ICOMOS Bureau.[96] By 1998 all nominated cultural sites were already assigned an expert mission, and the names of the experts undertaking the missions were published in the ICOMOS News.[97] Still, there were similar concerns about single-person dominance concerning the working methods of British archaeologist Henry Cleere, Léon Pressouyre's successor as the ICOMOS World Heritage coordinator from 1992–2002.[98]

There has been a recent emphasis – both by ICOMOS and the World Heritage Committee – on further transparency, credibility, high professional status and objectivity in the World Heritage evaluation procedures. In 2006 the World Heritage Working Group was established to coordinate ICOMOS' World Heritage work. More permanent staff has been recruited for the World Heritage Unit at the ICOMOS Secretariat, and to work as World Heritage advisors, replacing the former World Heritage coordinator. In addition, in January 2006 the ICOMOS Executive Committee adopted the "Policy for the implementation of the ICOMOS World Heritage mandate" (later amended in 2007, 2010 and again in 2012). This document aims at setting up standards for the World Heritage work and affiliations of experts engaged by ICOMOS. For instance, experts in any way involved in the preparation of national nomination files should not take part in the writing of ICOMOS evaluations. The objective of creating openness in the evaluation process is, however, somewhat contradicted by the strong conviction expressed in the document that both the World Heritage Working Group and the World Heritage Panel meetings should be in closed sessions.[99]

At present, once a new nomination has been sent to ICOMOS for evaluation, the procedure includes the following steps: initial examination of nomination dossiers at the ICOMOS World Heritage Unit; consultation with the scientific committees of ICOMOS and National Committees; possible consultation with other specialist bodies;[100] choosing the experts to be consulted for each nomination; compiling of experts' reports; producing a draft evaluation and recommendation based on these reports by the World Heritage advisor; presentation of the draft evaluation to the ICOMOS World Heritage Panel, consisting of ICOMOS Executive Committee members and yearly invited experts; and revision of the evaluation documents, if needed, based on the decisions made by the Panel. Each nomination is examined by two groups of experts: one that, on-desk and employing existing literature and documentation in consultation with chosen experts, attests to the outstanding universal value of the nominated site, and one that carries out an on-site evaluation mission, which assesses the existing conservation practices, management plans, authenticity, and integrity related to the site. When selecting experts for on-site missions, ICOMOS prefers professionals from the region where the nominated site is located, excluding those from the same country. In other words, the person making the assessment should have sufficient local knowledge but not too much engagement.[101] Two things should be pointed out: firstly, the definition of outstanding universal value continues to be dependent mainly on the on-desk review; the on-site missions should not assess the outstanding universal value but merely conduct a technical review of the site's condition in relation to management, authenticity and integrity. Secondly, the role of World Heritage advisors (five altogether at the moment) remains very central in the process. Despite the widely representative World Heritage Panel, which formulates the final statements of ICOMOS, there is only one advisor in each case who writes up and presents the draft report to the Panel. The same person also presents the final evaluation to the World Heritage Committee.

The ICOMOS evaluation procedures have surely gained more professional status over time. This can also be decoded from the format of the evaluation documents, for which, as pointed out in Chapter 1, there has been a significant expansion both in content and length. The proposed sites are today scrutinized from various angles and through cross-reference, even though there still seem to be variations between different ICOMOS evaluations in relation to the level and depth of the inquiry. The most striking overall feature of these documents, however, is their standardization.

When it comes to significance assessment, ICOMOS' (or IUCN's) expertise and objectivity, during the forty years of implementation of the Convention, have rarely been called into question. Natasha Affolder asserts that the active involvement of NGOs in the World Heritage work can be regarded "either as a sign of the democratic health of the regime, or, conversely, as an indicator of its anti-democratic nature, revealing the extent to which interest groups dominate global institutions."[102] The representatives of intergovernmental or non-governmental organizations, with similar objectives to those of ICOMOS, IUCN and ICCROM, may be considered in an advisory capacity within the World Heritage–related processes[103]; however, all non-state actors that might have an interest have not been given the same status as the official advisory bodies.[104]

Further transparency, credibility, high professional status and objectivity in the World Heritage evaluation process are obviously all favorable objectives. This discourse of objectivity and "apolitical universalism," as noted by Colin Long and Sophia Labadi, has an important function in the system: if the processes were entirely politicized, listing of new sites would be almost impossible.[105] The reverse of this discourse on objectivity, however, is the fading out of relativity in evaluating outstanding universal value, equally present in all value assessments. That the ICOMOS documents in the World Heritage language are called "evaluations," or even in some cases "technical evaluations," does not change their nature. Even though aiming for objectivity, they are *valuations*. The personal authorship of evaluation documents is always present. The selection of experts matters, as do their countries of origin, fields of expertise, commitments, or groups of reference. The ICOMOS World Heritage Panel is a conglomeration of people with different interests. It is also, interestingly, a group which has stemmed from surprisingly similar backgrounds: an audit report of 2010 indicated that in 2008, 14 out of the total of 25 ICOMOS Word Heritage Panel members were Europeans, three members came from Africa, three from Latin America and the Caribbean, five from Asia Pacific, and none from the Arab States. Nineteen representatives were men, six were women.[106] Still, the fact that ICOMOS, through the statements it makes, participates in the construction of meaning of places often remains invisible behind the curtain of objectivity.

As a remaining point, the relationship of ICOMOS to States Parties, on the one hand, and to the World Heritage Committee, on the other hand, in defining outstanding universal value deserves some comments. In the case of the former, ICOMOS always operates within the parameters laid out in the national nomination dossier with regard to the principle definition of outstanding universal value advocated therein. This means that ICOMOS cannot change the focus of the proposal completely in the course of its evaluation. If it rejects the foundation on which the nomination has been based but sees potential for another type of outstanding universal value, it may ask the state to re-nominate the site under this new focus.[107] At the textual level, it is also not difficult to find a correlation between national nomination dossiers and ICOMOS evaluations. In many cases ICOMOS has reproduced elements of the national nomination dossiers, especially those dealing with the history and description of the site, which in the ICOMOS documents often are summaries of longer descriptions presented by states. Still, even though relying partially on the national nomination texts, what ICOMOS chooses to represent is always its own choice, based on a conscious selection and valuation. Moreover, reflections in the evaluation documents concerning outstanding universal value and the criteria, authenticity, integrity and management, always represent independent work by ICOMOS. Thus, the ICOMOS evaluations on outstanding universal value should be read as combinations of interpretations of heritage as presented in national nomination dossiers, scholarly literature, on-site observation and ICOMOS' own valuation work. They should be read as heritage valuations.

Also, ICOMOS' and the World Heritage Committee's roles in the process of defining outstanding universal value continue to be close-knit. The principle of adhering to the Operational Guidelines and other Committee decisions and policies is supposed to guide the work of the advisory bodies.[108] The interchange between the two bodies takes place especially through the changing evaluation criteria on which ICOMOS bases its valuation. The criteria are decided on by the Committee but they have always been drafted in close cooperation with ICOMOS. At the same time, it is important to note that the valuations made by ICOMOS are not merely reproductions of the established criteria but involve their application, interpretation and more detailed articulation. Important places of interaction between the World Heritage Committee and ICOMOS with regard to the definition of outstanding universal value are also the actual Committee meetings, in which ICOMOS presents its evaluations. Recently, ICOMOS' definitions of outstanding universal value have become more openly contested by some of the World Heritage Committee members. For instance, in the 2011 Committee meeting the Delegation of Brazil raised the issue that ICOMOS had in its evaluation concerning the Historic Bridgetown and its Garrison (Barbados) made several factual errors, which, according to the delegate, "no doubt had substantial effect on the interpretation of the site values." ICOMOS proposed a deferral of the nomination, but the Committee decided to add Bridgetown to the World Heritage List without a delay.[109] The statement by the Delegation of Brazil interestingly presented the State Party, Barbados in this case, as the repository of knowledge concerning *intrinsic* outstanding universal value.

Notes

1 For the only exception, the Old City of Jerusalem, nominated by the Hashemite Kingdom of Jordan, see UNESCO, World Heritage Committee, First Extraordinary Session (Paris, 10–11 September 1981), Report of the Rapporteur, Paris, 30 September 1981, CC-81/CONF. 008/2 Rev., http://whc.unesco.org/archive/1981/cc-81-conf008-2reve.pdf.
2 For discussion, see James L. Hevia, "World Heritage, national culture, and the restoration of Chengde," *Positions* 9: 1 (2001). 219–243; Julie Scott, "World Heritage as a Model for Citizenship: the Case of Cyprus," *International Journal of Heritage Studies* 8: 2 (2002): 99–115; David Harrison, "Introduction," in *The Politics of World Heritage: Negotiating Tourism and Conservation*, ed. David Harrison and Michael Hitchcock (Clevedon: Channel View Publications, 2005); Bart J.M. van der Aa, *Preserving the Heritage of Humanity? Obtaining World Heritage Status and the Impacts of Listing*. Unpublished Ph.D. Thesis, University of Groeningen, 2005, 41–48; J. Ashworth and Bart J. M. van der Aa, "Strategy and Policy for the World Heritage Convention: Goals, Practices and Future Solutions," in *Managing World Heritage sites*, ed. Anna Leask and Alan Fyall (Amsterdam: Elsevier, 2006); Anna Leask, "World Heritage Site Designation," in *Managing World Heritage Sites*, ed. Anna Leask and Alan Fyall (Elsevier: Amsterdam, 2006), 8–9; Marc Askew, "The Magic List of Global Status. UNESCO, World Heritage and the Agendas of States," in *Heritage and Globalisation*, ed. Sophia Labadi and Colin Long (London: Routledge, 2010).
3 Jokilehto et al., *The World Heritage List: Filling the Gaps*; van der Aa, *Preserving*, 25; Sue Millar, "Stakeholders and Community Participation," in *Managing World Heritage Sites*, ed. Anna Leask and Alan Fyall (Elsevier: Amsterdam, 2006), 42.
4 Labadi, *UNESCO*, 20.
5 Moreover, the World Heritage Convention was drafted within the climate of recent decolonization, something which could also be seen in the constellation of its first signatories, including many newly independent Asian and African countries. For the new manner of relevance of UNESCO in the world of the 1970s, see J. P. Singh, *United Nations Educational, Scientific and Cultural Organization (UNESCO). Creating Norms for a Complex World* (New York: Routledge, 2011), 16–17, 91–92.
6 Titchen, *On the Construction*, 53–73.
7 Cameron and Rössler, *Many Voices*, 1–26.
8 Titchen, *On the Construction*, 14–35; Jokilehto, *A History*, 284; Fernando Valderrama, *A History of UNESCO* (Paris: UNESCO, 1995), 1–18.
9 Jokilehto, *A History*, 281–282; Dingli, Sandra M., "A plea for responsibility towards the common heritage of mankind," in *The Ethics of Archaeology: Philosophical Perspectives on Archaeological Practice*, ed. Chris Scarre and Geoffrey Scarre (Cambridge: Cambridge University Press, 2006), 222.
10 Swenson, *The Rise of Heritage*; Melanie Hall, "Introduction: towards World Heritage," in *Towards World Heritage. International Origins of the Preservation Movement 1870–1930*, ed. Melanie Hall (Aldershot: Ashgate, 2011), 2–4.
11 "Message from Mr. J. Torres Bodet the Director-General of UNESCO," *Museum* III: 1 (1950): 6.
12 Getty Conservation Institute, "Principles, Practice, and Process: A Discussion about Heritage Charters and Conventions," *The Getty Conservation Institute Newsletter* 19: 2 (2004), www.getty.edu/conservation/publications_resources/newsletters/19_2/dialogue.html.

13 Piero Gazzola, "Foreword," in *The Monument for the Man: Records of the II International Congress of Restoration*, Venezia, 25–31 Maggio 1964 (Padova: Marsilio, 1971), www. international.icomos.org/publications/hommeprein.pdf.

14 Waterton, Smith and Campbell, "The Utility of Discourse Analysis," 341; Matthew Hardy ed., *The Venice Charter Revisited: Modernism, Conservation and Tradition in the 21st Century* (Newcastle upon Tyne: Cambridge Scholars Publishing, 2011).

15 Harrison, *Heritage: Critical Approaches*, 27.

16 UNESCO, *Convention Concerning the Protection of World Cultural and Natural Heritage (World Heritage Convention)* (Paris: UNESCO, 1972), accessed July 12, 2015, http://whc.unesco.org/en/conventiontext/, Preamble, Article 11.

17 Askew, "The Magic List," 25.

18 Kathleen Adams, "The Politics of Heritage in Tana Toraja, Indonesia: Interplaying the Local and the Global," *Indonesia and the Malay World* 31: 89 (2003): 93.

19 Gibson and Pendlebury, "Introduction: Valuing Historic Environments," 6–7.

20 *World Heritage Convention* 1972, Preamble, Articles 3, 4, 6, 7. See also Natasha Affolder, "Democratising or Demonising the World Heritage Convention?" *Victoria University of Wellington Law Review* 38: 2 (2007): 342.

21 *International Charter for the Conservation and Restoration of Monuments and Sites (The Venice Charter)* (Venice: The Second Congress of Architects and Specialists of Historic Buildings, 1964), accessed October 15, 2015, www.icomos.org/charters/venice_e.pdf, Article 1.

22 See, for example, UNESCO, General Conference, Thirteenth Session (20 October – 19 November 1964), Item 15.3.4 of the Provisional Agenda, Report on Measures for the Preservation of Monuments of Historical or Artistic Value, 13 C/PRG/15, Paris, 16 June 1964, http://whc.unesco.org/archive/1964/13c-prg-15e.pdfA.

23 UNESCO, Meeting of Experts to Establish an Intergovernmental System for Protection of Monuments, Groups of Buildings and Sites of Universal Interest (21–25 July 1969), Final report, SHC/MD/4, Paris, 10 November 1969, p. 28–29, http://whc.unesco.org/archive/1969/shc-md-4e.pdf.

24 *World Heritage Convention* 1972, Article 1.

25 Titchen, *On the Construction*, 90–94.

26 Jokilehto, Jukka et al., *What is OUV?*.

27 Henry Cleere, "The concept of 'outstanding universal value' in the World Heritage Convention," *Conservation and Management of Archaeological Sites*, 1: 4 (1996): 232. See also Henry Cleere, "The uneasy bedfellows: universality and cultural heritage," in *Destruction and Conservation of Cultural Property*, ed. Robert Layton, Peter G. Stone and Julian Thomas, (London / New York: Routledge, 2001), 25.

28 Tim Winter, "Heritage tourism: The dawn of a new era?" in *Heritage and Globalisation*, ed. Sophia Labadi and Colin Long (Routledge: London, 2010), 117–129.

29 Lynn Meskell, "Negative heritage and past mastering in archaeology." *Anthropological Quarterly* 75: 3 (2002): 568. See also Byrne, "Western hegemony," 269–276.

30 UNESCO, Informal Consultation of Intergovernmental and Non-Governmental Organizations on the Implementation of the Convention concerning the Protection of the World Cultural and Natural Heritage (Morges, 19–20 May 1976), CC-76/WS/25, p. 2. UNESCO, World Heritage Centre, WH Committee & Bureau, Working Documents from 1976 to 1987 (CD in author's possession).

31 UNESCO, World Heritage Committee, First Session (UNESCO, Paris, 27 June – 1 July 1977), Issues Arising in Connection with the Implementation of the World Heritage Convention, CC-77/CONF.001/4, Paris, 9 June 1977, p. 6, http://whc.unesco.org/archive/1977/cc-77-conf001-4e.pdf.

32 Jokilehto, Jukka et al., *What is OUV?*, 48.

33 UNESCO, Bureau of the World Heritage Committee, Twenty-Second Session (Paris, 22–27 June 1998), Information Document: Report of the World Heritage Global Strategy Natural and Cultural Heritage Expert Meeting, 25 to 29 March 1998, Theatre

Institute, Amsterdam, The Netherlands, WHC-98/CONF.201/INF.9, Paris, 3 June 1998, p. 15, http://whc.unesco.org/archive/1998/whc-98-conf201-inf9e.pdf.
34 Jokilehto, *A History*, 295–296.
35 Michael Petzet, "Introduction," in *What is OUV? Defining the Outstanding Universal Value of Cultural World Heritage Properties*, ed. Jukka Jokilehto et al. Monuments and sites XVI (Berlin: Hendrik Bäßler Verlag, 2008), 8.
36 UNESCO, World Heritage Committee, Twenty-Ninth Session (Durban, South Africa, 10–17 July 2005), INF.9B: Keynote Speech by Ms Christina Cameron and Presentations by the World Heritage Centre and the Advisory Bodies, WHC-05/29.COM/INF.9B, Paris, 15 June 2005, p. 2, http://whc.unesco.org/archive/2005/whc05-29com-inf09Be.pdf.
37 Labadi, *UNESCO*, 54.
38 For an early statement regarding the exclusive nature of the World Heritage List as a guarantee of its credibility, see, for example, UNESCO, World Heritage Committee, First Session (Paris, 27 June – 1 July 1977), Final Report, CC-77/CONF.001/9, Paris, 17 October 1977, p. 3–4, http://whc.unesco.org/archive/1977/cc-77-conf001-9_en.pdf.
39 Ashworth and van der Aa, "Strategy and Policy," 150.
40 See, for example, UNESCO, World Heritage Committee, Fifth Session (Sydney, 26–30 October 1981), Report of Rapporteur, CC-81/CONF.003/6, Paris, 5 January 1982, p. 8, http://whc.unesco.org/archive/1981/cc-81-conf003-6_e.pdf.
41 UNESCO, World Heritage Committee, Third Session (Cairo and Luxor, 22–26 October 1979), Report of Rapporteur, CC-79/CONF.003/13, Paris, 30 November 1979, p. 2, http://whc.unesco.org/archive/1979/cc-79-conf003-13e.pdf.
42 UNESCO, World Heritage Committee, Eighteenth Session (Phuket, Thailand, 12–17 December 1994), Information Note: Nara Document on Authenticity. Experts meeting, 1–6 November 1994, WHC-94/CONF.003/INF.008, 21 November 1994, http://whc.unesco.org/archive/1994/whc-94-conf003-inf8e.pdf; UNESCO, World Heritage Committee, Twenty-Ninth Session (Durban, South Africa, 10–17 July 2005), Assessment of the Conclusions and Recommendations of the Special Meeting of Experts (Kazan, Russian Federation, 6–9 April 2005), WHC-05/29.COM/9, Paris, 15 June 2005, p. 3, http://whc.unesco.org/archive/2005/whc05-29com-09e.pdf.
43 See also Labadi, *UNESCO*, 57.
44 UNESCO, Operational Guidelines for the Implementation of the World Heritage Convention, 8 July 2015, Paragraphs 116, 192b. Hereafter the versions of the document are referred to as Operational Guidelines. The different stages in the development of this document can be found at http://whc.unesco.org/en/guidelines/.
45 UNESCO, World Heritage Committee, Twenty-Ninth Session (Durban, South Africa, 10–17 July 2005), Assessment of the Conclusions and Recommendations of the Special Meeting of Experts (Kazan, Russian Federation, 6–9 April 2005), WHC-05/29. COM/9, Paris, 15 June 2005, p. 2, http://whc.unesco.org/archive/2005/whc05-29com-09e.pdf. See also Kishore Rao, "A new paradigm for the identification, nomination and inscription of properties on the World Heritage List," *International Journal of Heritage Studies* 16: 3 (2010): 161–172; or Jade Tabet, *Review of ICOMOS' working methods and procedures for the evaluation of cultural and mixed properties* (Paris: ICOMOS, 2010), 27, accessed October 15, 2015, www.icomos.org/world_heritage/WH_Committee_34th_session_Brasilia/JT_Final_report_en.pdf.
46 Gibson and Pendlebury, "Introduction: Valuing Historic Environments," 7.
47 UNESCO, World Heritage Committee, Twenty-Ninth Session (Durban, South Africa, 10–17 July 2005), Assessment of the Conclusions and Recommendations of the Special Meeting of Experts (Kazan, Russian Federation, 6–9 April 2005), WHC-05/29.COM/9, Paris, 15 June 2005, p. 3, http://whc.unesco.org/archive/2005/whc05-29com-09e.pdf.
48 Petzet, "Introduction," 8; Cleere, "The concept," 231; UNESCO, World Heritage Committee, Seventh Session (Florence, Italy, 5–9 December 1983), Speech by Mr. Michel Parent, Chairman of ICOMOS, during the Seventh Session of the Bureau of

the World Heritage Committee (Paris, 27–30 June 1983), SC/83/CONF.009/INF.2, Paris, 1 September 1983, p. 2, http://whc.unesco.org/archive/1983/sc-83-conf009-inf2e.pdf.

49 Cleere, "The Uneasy Bedfellows," 23.

50 Avarami, Mason and de la Torre, *Values and Heritage Conservation*, 8.

51 *World Heritage Convention* 1972, Article 1.

52 Petzet, "Introduction," 7.

53 UNESCO, World Heritage Committee, First Session (Paris, 27 June – 1 July 1977), Final Report, CC-77/CONF.001/9, Paris, 17 October 1977, p. 4, http://whc.unesco.org/archive/1977/cc-77-conf001-9_en.pdf.

54 See also Labadi, *UNESCO*, 34–35.

55 Cameron and Rössler, *Many Voices*, 35.

56 UNESCO, World Heritage Committee, Third Session (Cairo and Luxor, 22–26 October 1979), Report of Rapporteur, CC-79/CONF.003/13, Paris, 30 November 1979, p. 9, http://whc.unesco.org/archive/1979/cc-79-conf003-13e.pdf.

57 UNESCO, World Heritage Committee, Third Session (Luxor, Arab Republic of Egypt, 23–27 October 1979), Comparative Study of Nominations and Criteria for World Cultural Heritage. Report by Mr. M. Parent, Paris, 20 September 1979. CC-79/CONF.003/11 Annex, p. 22, 24, http://whc.unesco.org/archive/1979/cc-79-conf003-11e.pdf.

58 Labadi, *UNESCO*, 35.

59 Sophia Labadi, "A Review of the Global Strategy for a Balanced, Representative and Credible World Heritage List 1994–2004," *Conservation and Management of Archaeological Sites* 7: 2 (2005), 95.

60 Operational Guidelines, 8 July 2015, Article 77.

61 Mason, "Assessing Values," 5–30.

62 Alois Riegl, "The Modern Cult of Monuments: Its Character and Its Origins," *Oppositions* 25 (1982) [1903]: 20–51.

63 Bernard M. Feilden and Jukka Jokilehto, *Management Guidelines for World Cultural Heritage Sites*. 2nd cdn (Rome: ICCROM, 1998), 18–21.

64 Derek Worthing and Stephen Bond, *Managing Built Heritage: The Role of Cultural Significance* (Oxford: Blackwell Publishing, 2008), 66, 68–69. Rebecca Madgin contributes to the discussion on contemporary heritage values from the point of view of how these values were ascribed in the actual decisions to retain or demolish old urban buildings. Rebecca Madgin, *Heritage, Culture and Conservation: Managing the Urban Renaissance* (Saarbrücken: Verlag Dr. Müller, 2009).

65 Mason, "Assessing Values," 10–12.

66 Feilden and Jokilehto, *Management Guidelines*, 19.

67 Mason, "Assessing Values," 11.

68 Ibid., 13. The intrinsic value argument in heritage conservation presumes that by proving the material authenticity of a truly old heritage material, historical value is established.

69 Australia ICOMOS, *The Burra Charter. The Australia ICOMOS Charter for Places of Cultural Significance* (Victoria: Australia ICOMOS, 1999), accessed October 15, 2015, http://australia.icomos.org/wp-content/uploads/BURRA_CHARTER.pdf, 12; Mason, "Assessing Values," 11.

70 Worthing and Bond, *Managing Built Heritage*, 62.

71 Walter Lipe, "Value and Meaning in Cultural Resources," in *Approaches to the Archaeological Heritage. A Comparative Study of World Cultural Resource Management Systems*, ed. Henry Cleere (Cambridge: Cambridge University Press, 1984), 7. Moreover, Lipe notes how heritage can be aesthetically valued without knowledge about its past, potentially leading to a use that trivializes that history.

72 *The Burra Charter* 1999, 12.

73 Mason, "Assessing Values," 12; Worthing and Bond, *Managing Built Heritage*, 68.

74 Mason, "Assessing Values," 11.

75 Chris Johnston, *What is Social Value?* (Australian Heritage Commission: Canberra, 1994), 5. Op. cit. in Gibson, "Cultural Landscapes and Identity," 73.
76 Mason, "Assessing Values," 12.
77 Munasinghe, *Urban Conservation and City Life*, 32, 37, 237.
78 Petzet, "Introduction," 9.
79 *The Burra Charter* 1999, 2.
80 Council of Europe Framework Convention on the Value of Cultural Heritage for Society 2005, Faro 27.X.2005, accessed October 15, 2015, www.coe.int/en/web/conventions/full-list/-/conventions/rms/0900001680083746.
81 Sarah M. Titchen, "On the Construction of 'Outstanding Universal Value.' Some Comments on the Implementation of the 1972 UNESCO World Heritage Convention," *Conservation and Management of Archaeological Sites*, 1: 4 (1996): 237.
82 Feilden and Jokilehto, *Management Guidelines*, 18–21.
83 Smith, *Uses of Heritage*, 88.
84 For an early description concerning the role of ICOMOS and IUCN see UNESCO, World Heritage Committee, Fifth Session (Sydney, 26–30 October 1981), Report of Rapporteur, CC-81/CONF.003/6, Paris, 5 January 1982, p. 28, http://whc.unesco.org/archive/1981/cc-81-conf003-6_e.pdf.
85 ICOMOS, Paris, 7 June 1979, Ernst Connally, Secretary General to Mr. Firouz Bagerzadeh, Chairman, World Heritage Committee. World Heritage Centre, World Heritage Sites, Nomination files (1978–1997), The Americas (CD in author's possession).
86 Cameron and Rössler, *Many Voices*, 188–190.
87 Francois LeBlanc, "An Inside View of the Convention," *Monumentum*, Special Issue, 1984, 21.
88 Ibid., 21–22.
89 Cameron and Rössler, *Many Voices*, 187.
90 LeBlanc, "An Inside View," 17–32; UNESCO, World Heritage Committee, Seventh Session (Florence, Italy, 5–9 December 1983), Speech by Mr. Michel Parent, Chairman of ICOMOS, during the Seventh Session of the Bureau of the World Heritage Committee (Paris, 27–30 June 1983), SC/83/CONF.009/INF.2, Paris, 1 September 1983, http://whc.unesco.org/archive/1983/sc-83-conf009-inf2e.pdf; Léon Pressouyre, *The World Heritage Convention, Twenty Years Later* (Paris: UNESCO, 1996), 39.
91 UNESCO, World Heritage Committee, Seventh Session (Florence, Italy, 5–9 December 1983), Report of Rapporteur, C/83/CONF.009/8, Paris, January 1984, Annex II: Address by the Outgoing Chairman of the World Heritage Committee, Professor Ralph Slatyer, http://whc.unesco.org/archive/1983/sc-83-conf009-8e.pdf.
92 UNESCO, World Heritage Committee, Seventh Session (Florence, Italy, 5–9 December 1983), Speech by Mr. Michel Parent, Chairman of ICOMOS, during the Seventh Session of the Bureau of the World Heritage Committee (Paris, 27–30 June 1983), SC/83/CONF.009/INF.2, Paris, 1 September 1983, p. 2, http://whc.unesco.org/archive/1983/sc-83-conf009-inf2e.pdf.
93 Ibid.
94 Cameron and Rössler, *Many Voices*, 191.
95 Ronald Silva, "Report of the President of ICOMOS, 1990–1995," *Thirty years of ICOMOS. Scientific Journal* 5 (1995), 119.
96 ICOMOS, No 614, October 1994 (Safranbolu, Turkey).
97 *ICOMOS News* 8:1 (1998), 7–8.
98 Cameron and Rössler, *Many Voices*, 193–4.
99 Policy for the Implementation of the ICOMOS World Heritage Mandate. ICOMOS Executive Committee 17 January 2006 and as amended in November 2007, in October 2010 and in October 2012, accessed October 15, 2015, www.international.icomos.org/world_heritage/ICOMOS_WH_Policy_paper_201011_EN.pdf.

100 An audit report from 2010 shows that ICOMOS has been reluctant to consult non-member experts. For instance, in 2006–2009 out of the 22 historic town nominations examined by ICOMOS, only in two cases was an external expert consulted. Tabet, *Review of ICOMOS' working methods*, 22–23.

101 ICOMOS, The Role of ICOMOS in the World Heritage Convention, accessed March 9, 2016, www.icomos.org/en/what-we-do/image-what-we-do/268-he-role-of-icomos-in-the-world-heritage-convention?showall=&limitstart=; Tabet, *Review of ICOMOS' working methods*.

102 Affolder, "Democratising," 358–59.

103 *World Heritage Convention* 1972, Article 8.

104 Affolder, "Democratising," 359.

105 Colin Long and Sophia Labadi, "Introduction," in *Heritage and Globalisation*, ed. Sophia Labadi and Colin Long (London: Routledge, 2010), 8–9.

106 Tabet, *Review of ICOMOS' working methods*, 28–29.

107 See, for example, ICOMOS evaluation for the nomination of the World Heritage property, "Provins," France, No 873, October 1998. World Heritage Centre, World Heritage Sites, Rejected Nominations (CD in author's possession).

108 Operational Guidelines, November 2011, Paragraph 148.

109 UNESCO, World Heritage Committee, Thirty-Fifth Session, Paris, UNESCO Headquarters (19–29 June 2011), Summary Record, WHC-11/35.COM.INF.20, 205–206, http://whc.unesco.org/archive/2011/whc11-35com-inf20.pdf.

3   Catching up with "the spirit of the moment"

The World Heritage Convention was on several counts a progressive opening at the time of its creation. In contrast, the early implementation of the Convention in the 1980s may be seen as a period when many of the good intentions outlined in the Convention were set aside in favor of an exercise in listing new sites under the World Heritage heading. Even though not as black and white as this, against the background of the 1980s the early 1990s presents itself as a period of significant reorientation and self-reflection in the history of the implementation of the Convention. Since the 1970s, the perception of the concept of cultural heritage had undergone significant widening in professional conservationist and academic circles (see Chapter 1). By the end of the 1980s many of the adherents of the World Heritage idea felt that the implementation of the Convention should catch up with "the spirit of the moment" to be found elsewhere in the professional heritage field, especially with regard to "the relationships of man to his environment," and emerging themes such as anthropized landscapes or vernacular architecture, as summed up by the representative of ICOMOS at the 1989 World Heritage Committee meeting.[1]

The early 1990s reorientation, together with its practical and discursive limitations, is the focus of this chapter, which addresses five processes in particular. These include launching the Global Strategy for a more representative World Heritage List; introducing a new category of cultural landscapes for World Heritage nominations; widening the Western-based concept of authenticity, considered essential in the process of defining outstanding universal value; bringing the notion of intangible heritage into the debates on World Heritage; and assigning a larger role to local communities in defining World Heritage. The closer look at these initiatives and their strengths and weaknesses is warranted because one of the main objectives of this book is to reflect on the scope of the reorientation in relation to ICOMOS discourses regarding World Heritage cities. The last section of this chapter already touches upon this question, as it discusses the role of urban heritage as part of the definition of a 'new era' in the implementation of the World Heritage Convention.

In many ways, the 'state of affairs' for urban heritage within the World Heritage framework in the early 1990s was defined by a specific set of guidelines, adopted by the World Heritage Committee in 1984 on the basis of an ICOMOS expert meeting and included in the 1987 Operational Guidelines. On this occasion, cities eligible for inclusion in the World Heritage List were grouped into three categories: the no longer inhabited towns, which "generally satisfy the general criterion of authenticity and can be easily managed"; the still inhabited historic towns "which, by their very nature, have developed and will continue to develop under the influence of socioeconomic and cultural change, a situation that renders the assessment of their authenticity more difficult and any conservation policy more problematical"; and the twentieth-century "new towns," the authenticity of which was considered "undeniable," but the future of which was seen as "unclear because their development cannot be controlled."[2]

Broadening concepts of World Heritage

Globalizing the World Heritage Convention

The reorientation of the early 1990s culminated in one document in particular, the *Global Strategy for the Implementation of the World Heritage List*, drafted at a meeting of experts and approved by the World Heritage Committee in 1994. By that time it had become evident for those involved in the daily World Heritage work that the construction of the World Heritage List had so far been based on what was considered a narrow concept of cultural heritage. This led to the replacement of the rigid and mechanistic classification based on a traditional chronological, art historical and civilizations approach, the so-called Global Study,[3] in favor of a more dynamic and thematic approach of Global Strategy. At the heart of the ongoing Global Strategy process is the representativeness of the World Heritage List. As such, it continues the debates concerning balance and imbalance that have been on the Committee's agenda since its first meeting in 1977.[4]

The Expert Meeting on the Global Strategy of 1994 pointed out severe geographic, chronological and thematic imbalances in the World Heritage List. The detected aspects of imbalance included an over-representation of Europe compared with other cultural regions; of religious buildings, historic towns and monumental and elitist architecture as opposed to other types of cultural heritage and more vernacular expressions; of Christian sites as opposed to sites associated with other religions; and of historical periods compared with pre-history and the twentieth century. The overall conclusion was that the World Heritage List failed to reflect "living cultures, ethnographic and archaeological landscapes, and many of the broad areas of human activity" of outstanding universal value. Consequently the adoption of a more anthropological, multi-functional, universal and contextual view towards cultural heritage was urged. In order to rectify the imbalances in the World Heritage List as it then stood, the Expert Meeting identified two broad frameworks, "Human coexistence with the land" and "Human beings in society." The identification of under-represented themes was to be carried out in regional meetings.[5]

The Global Strategy challenged many of the foundations of the authorized approach to heritage within the World Heritage regime. It called for a change in the system. Emma Waterton, Laurajane Smith and Gary Campbell, in their analysis of the 1999 Burra Charter version, stress that, while sincerely attempting to open up cultural heritage management processes towards community inclusion, this Australian landmark heritage charter nevertheless re-established the authority of conservation professionals and the authorized discourses of heritage.[6] The scope of change brought about by the Global Strategy will be further explored in the following chapters; here we should note that while the drafting of the Global Strategy was marked by a similarly sincere effort at wider representation and at opening up new discussions, as motivated the Burra Charter, it too continued to reproduce established discourses.

This may be pointed out in relation to the ultimate clinging of the Global Strategy to a modernist concept of heritage. For example, despite the desire to depart from the typological classification approach of the Global Study, and while applying wider chronological-regional and thematic frameworks, typological considerations have continued to be part of the Global Strategy initiative. Also testifying to its clinging to a modernist concept of heritage are the continuing objective of attaining a global view of all heritage of the world and the practice of categorization itself, even though it is now based on a wider array of variables than previously. The effort to fill in the "gaps" on the World Heritage map has been advanced with ever more effort since 1994.[7] The filling-of-gaps activity has a dual modernist meaning. On the one hand, it implies that by identifying imbalances, the World Heritage List could someday be fully representative and complete. On the other hand, and in complete contrast to this position, it shows itself to be an endless and self-sustaining exercise. It is always possible to identify new imbalances and little-represented themes. Thus, the filling-the-gaps exercise promises to legitimize the continuity of the World Heritage identification work until far into the future.

Also worth considering is the fact that the Global Strategy continued to place strong faith in "objective criteria and operational procedures" in producing selections that are "truly relevant."[8] Thus the Global Strategy exercise is largely in tune with the overall paradox in the context of World Heritage valuation where it was considered that the valuation of cultural heritage should be carried out according to culture-specific standards but where outstanding universal value is assumed to be somehow intrinsic to a heritage place. In other words, even if there has been a shift from modernist towards postmodernist understanding of cultural heritage, as Julian Smith observes, the latter, too, aims to define cultural relativity within an ongoing universalist framework.[9]

In the intersection of culture and nature: cultural landscapes

The second conceptual reassessment that warrants discussion relates to the concept of cultural landscape. The reference to the "combined works of nature and man of outstanding universal value" in the World Heritage Convention puzzled the World Heritage Committee during the entire 1980s.[10] In 1992, after years of preparation and in a response to initiatives from lower-level actors,[11] the Committee adopted cultural landscape as a new nomination category to be applied in World

Heritage listing.[12] This made the World Heritage Convention the first international legal instrument to speak about cultural landscapes and their protection.[13] At the heart of the concept was, and still is, the acknowledgement of interactions between culture and nature, between people and the environment, and between tangible and intangible manifestations of heritage. Cultural landscapes are understood to be "illustrative of the evolution of human society and settlement over time, under the influence of the physical constraints and/or opportunities presented by their natural environment and of successive social, economic and cultural forces, both external and internal."[14] Along with Global Strategy conclusions, this definition opened up the concept of outstanding universal value to the idea of regional representativeness.

The Committee further identified three (or actually four) main types of this new type of heritage. The first category, *clearly defined landscapes*, refers to landscapes created intentionally by man, i.e. parks and gardens. The second category, *associative cultural landscapes*, involves landscapes, the outstanding universal value of which is justifiable by virtue of the powerful religious, artistic or cultural associations of the natural element rather than material cultural evidence, which may even be absent.[15] If one accepts that cultural landscapes are something experienced rather than observed and that they "exist in imagination, although in relationship to a specific place,"[16] all cultural landscapes actually fall into the category of associative landscapes.

The landscapes forming the third category, *organically evolved landscapes*, are considered to be those resulting from an initial social, economic, administrative and/or religious imperative and having developed their present form by association with, and in response to, their natural environment. These include, on the one hand, *relict landscapes*, in which evolutionary processes have come to an end at some time in the past, and, on the other hand, *continuing landscapes*, which retain an active social role in contemporary society closely associated with a traditional way of life, and in which the evolutionary process is still in progress. The latter category, most often used in the subsequent cultural landscape nominations,[17] has also been the most difficult to define and manage for World Heritage purposes. Henry Cleere suggested in 1995 that "organically evolved continuing landscapes cannot be 'frozen' in their state at the time when they are inscribed on the World Heritage List because, paradoxically, if so, they would not qualify as continuing landscapes any longer. Consequently, the insertion of modern elements into these landscapes should be allowed."[18] The challenging potential of organically evolved continuing landscapes thus relates to the question of how to combine retaining a close association with a traditional way of life with allowing a continuing process of change.

Reconsidering the Western approach to authenticity

To be included in the World Heritage List, cultural heritage sites must meet "the test of authenticity," together with meeting one or more of the valuation criteria discussed in Chapter 2. In the World Heritage system, authenticity is intended to be a kind of qualifying condition and a control concept that should be reflected on in relation to the outstanding universal value that has been identified based on the criteria.[19] The difference between heritage values and authenticity is that while the former are seen subject to social and historical construction, the latter is seen to either exist, or to not exist, and impossible to restore once lost.[20] However, this uncompromising nature of authenticity is complicated by knowing that the reflection points against which authenticity is evaluated at each time also change. This obviously leads to a situation where a heritage resource, perhaps earlier considered inauthentic, might, as the points of reflection change, be considered authentic.

The word 'authenticity' can be understood in multiple ways, for example as authority, genuineness, original state, realness, reliability, truthfulness or credibility. Also, what is authentic is always defined in relation to what is non-authentic – copied, pretended, counterfeited, forged, or without authority.[21] Even what little has so far been said about authenticity makes it clear that one is dealing with a concept which most likely can never be unambiguously defined. All the same, "the test of authenticity" has always been considered essential to the functioning of the system by most World Heritage practitioners. Even those who have emphasized the culturally dependent nature of the concept, and its Western origin, have nevertheless believed in its universal appeal.[22]

Being part of the regularly revised Operational Guidelines has enabled an evolutionary approach to authenticity as new ideas emerge. For a long time, however, this remained largely a theoretical possibility, since very few changes to the wording of the concept were actually initiated. For an early conceptualization of authenticity within the World Heritage framework one should look at the definition by an expert group in 1977: "the property should meet the test of authenticity in design, materials, workmanship and setting; authenticity does not limit consideration to original form and function but includes all subsequent modifications and additions, over the course of time, which in themselves possess artistic or historical values."[23] This statement may be read in two ways: firstly, as an effort to establish scientific measures for the "test of authenticity," and secondly, as an effort at dissociation from the most restricted interpretation of authenticity as the preservation of the original form. As noted by one member of the World Heritage Committee, being respective of a building's later modifications was considered "progressive authenticity."[24] The World Heritage Committee, however, decided to delete the reference to subsequent modifications from the 1980 version of the Operational Guidelines.[25] At an earlier date the Committee had already replaced the word "function" by the word "structure" in its definition of authenticity.[26] Whether these changes meant that the Committee adopted, without further consideration, later modifications and new uses as part of World Heritage sites, or hoped to avoid discussing the issue further, remains unclear from the Committee minutes. Nevertheless, it is clear that these changes narrowed the focus of the authenticity concept for over a decade.

Even though the question of later modification was important when considering the condition of authenticity, the main intention in requiring it was to argue against reconstruction and historical falsification.[27] The well-known test for the World Heritage Convention in this sense was Poland's nomination of the Historic Centre of Warsaw, heavily reconstructed after its destruction in World War II. ICOMOS supported the nomination, but required further expert opinion on its authenticity.[28] Among the members of the World Heritage Bureau, opinion on authenticity was divided.[29] The question was, as voiced by the President of ICOMOS, Michel Parent, "can a haphazard 19th Century or a systematic 20th Century reconstruction be justified for inclusion on grounds, not of Art but of History?"[30] For Warsaw the answer was positive. It was included in the World Heritage List as an exceptionally well implemented example of reconstruction associated with events of considerable historical significance, and because of its symbolic role in the patriotic feeling among the Polish people. It was, nevertheless, underlined that Warsaw represented an exception, and that no other reconstructed properties would be inscribed in the future.[31] This was in line with the Committee's ideas at that moment, and also later, that artistic value could be treated as a series within the World Heritage List, whereas "sites representing the positive and negative sides of human history will only be invested with real force if we make the most remarkable into unique symbols, each one standing for the whole series of similar events."[32]

The statement regarding Warsaw by the Committee suggested that 'non-authentic' would equal 'reconstructed,' and that, 'reconstructed,' apart from Warsaw, could not be World Heritage. To clarify its position towards reconstructions, the World Heritage Committee, in 1980, included in the Operational Guidelines a sentence noting that "reconstruction is only acceptable if it is carried out on the basis of complete and detailed documentation on the original and to no extent on conjecture."[33] This wording justified Warsaw's inscription, but in fact also left the door open for other reconstructions. In 1983 Michel Parent further explained that it was "not so much reconstitution that is condemned in principle as erroneous or fanciful restoration."[34]

The effort to make the World Heritage Convention more relevant to the diversity of world cultures in the early 1990s also led to the critical evaluation of the condition of authenticity.[35] In fact, the most extensive discussions of the 1990s in the whole professional cultural heritage field involved authenticity. With its profound reflection on the concept, the World Heritage community was in the vanguard of those professional debates.[36] The choice of venue for a conference elaborating on authenticity in the World Heritage context in 1994, Nara, Japan, outside Europe and in a country with a unique understanding of cultural heritage, served as a symbolic act towards inclusion of non-Western cultures.[37] The Nara Document on Authenticity, adopted by the conference, stressed that the basis for assessing authenticity was knowledge of the information sources (form and design, materials and substance, use and function, traditions and techniques, location and setting, spirit and feeling) concerning heritage values, "in relation to original and subsequent characteristics of the cultural heritage." Furthermore, it noted that the values attributed to cultural heritage, credibility of information sources and thus also the judgement on authenticity should be considered within the cultural contexts to which they belong. Authenticity was presented as a culture-specific and relative concept. Still, the Nara Document maintained the importance of defining authenticity through a scientific process.[38]

It is not difficult to see that a significant change in discourse took place during the mid-1990s. Allowing flexible interpretations of authenticity would allow inscription of non-Western sites that previously could not be considered. This notwithstanding, the relationship with regard to the concept of authenticity remained ambiguous after Nara. For one thing, the change in discourse was presumably too abrupt for some members of the World Heritage Committee and some States Parties,[39] which can be seen from the slow pace at which the Nara ideas became part of the Operational Guidelines. Only in the 2005 version of the document were attributes such as use and function, traditions, language and spirit and feeling added to the authenticity matrix to be used for evaluating World Heritage sites.[40]

A further example of ambiguous attitudes towards the concept of authenticity that prevailed after the Nara Conference is shown in the 1998 *Management Guidelines for World Cultural Heritage Sites*. While briefly referring to the recent Nara Document, and noting the conceptual variations in understanding authenticity, these guidelines made no actual reference to any other aspects of authenticity, except for those related to "material, workmanship, design and setting."[41] In terms of the relationship between different authenticities, material authenticity was seen as superior to other authenticities, for example functional authenticity: "Where applicable, a heritage resource should be allowed to continue to serve its traditional function, insofar as this does not cause damage to its historical integrity."[42] This brings me to my final consideration with regard to the discussions in Nara: the relationship between different authenticities in a situation where they stand in competition and in conflict with each other remained under-discussed. Making such a choice and prioritization, however, is often necessary in actual practice; for example, if truly emphasized, functional and spiritual authenticity often conflict with the objectives of material authenticity.

Intangible heritage

One obvious conclusion based on the definitions of heritage given in Chapter 1 – heritage as a discourse, as a process and as something that is utterly based on contemporary valuation – is that all heritage, even the most material kind, is intangible.[43] Heritage professionals, however, have traditionally drawn a stark dividing line between the two categories of tangible and intangible. This also holds true for UNESCO, which, in parallel with its preoccupation with monuments, groups of buildings and sites, has had a long-term interest in intangible heritage.[44] The 1990s experienced intensification of policy in this area as well. In 1993 the organization initiated the Living Human Treasures Programme,[45] and a few years later, in 1998, the Proclamation of Masterpieces of the Oral and Intangible Heritage of Humanity, a counterpart to the World Heritage List in the area of intangible cultural heritage.[46]

The most significant expression of UNESCO's recent commitment with respect to intangible heritage is the *Convention for Safeguarding the Intangible Cultural Heritage*, adopted in 2003 and henceforth to be referred to as the 2003 Convention. The 2003 Convention was significantly inspired by the Japanese heritage paradigm of *mukei bunkazia* which focuses on safeguarding the traditional processes of transmission of skills. As Chiara Bortolotto shows, the present international concern for intangible heritage, concurrent with contemporary anthropological theories, was, to a significant extent, exported from Japan and successfully turned into a global paradigm, suggesting that in the case of the 2003 Convention globalization did not equate to Westernization.[47] Moreover, Bortolotto relates the success of non-Western heritage models in UNESCO, on the one hand, to Japan joining the World Heritage Convention in 1992, and, on the other hand, to a Japanese directorship of UNESCO by Koïchiro Matsuura, from 1999 onwards. After the withdrawal of the USA from UNESCO in 1984, Japan, "uncomfortable with UNESCO heritage criteria," was the main contributor to the UNESCO extra-budgetary fund.[48] This example shows how UNESCO has been not just a globalizing agent from above, but open to redefinition of its concepts from below, especially from powerful member states.[49]

The Masterpieces Declaration and the 2003 Convention undeniably constitute the most serious efforts at transforming the international heritage paradigm thus far supported by UNESCO, marking a visible change in the international heritage discourse.[50] Both these initiatives were based on the realization that the World Heritage Convention was insufficient with regard to acknowledging non-material and even ordinary expressions of cultural heritage, and that complementary approaches were needed.[51] The Preamble to the 2003 Convention also recognizes the "deep-seated interdependence" between intangible and tangible heritage, and a significant role is assigned to communities as bearers of heritage.[52] Accordingly, the subjective heritage values of communities, instead of being judged by externally defined criteria, are treated as the basis for defining intangible cultural heritage. So is continuous change. Finally, while the Masterpieces Proclamation still referred to "outstanding value," the 2003 Convention discarded this notion as universalizing, and susceptible to creating unnecessary hierarchies within the system. Instead, and reflecting the cultural relativist thought behind it, the list that was established based on the 2003 Convention was titled the Representative List of the Intangible Cultural Heritage.[53]

One important reservation to the above described intangible heritage paradigm shift stems from the universalizing nature of all international standard-setting instruments, whatever their founding principles may be. The principles of cultural relativism, cultural diversity and empowerment of grassroots communities, in the definition of heritage dealt with in the 2003 Convention, are significantly undermined by the creation of a uniform standard-setting framework and the state-level validation process required for heritage authorization in this system.[54] Another reservation relates to the tangible-intangible division of cultural heritage. By definition, their co-existence is acknowledged; however, the existence of a dual system of heritage designation within the UNESCO framework serves to underline the distinction between the two. The existence of the dual system also becomes an end in itself – once created it needs to be justified.[55] In this mindset the World Heritage List continues to be the tangible list. The existence of a separate list for intangible heritage may thus actually justify the subordinate role given to intangible heritage aspects as part of the conceptualization of cultural heritage within the World Heritage system.

The third reservation, which is based on the current dominant Western value system, is that the World Heritage List is deemed to command more authority than its counterpart for intangible heritage, the Representative List of the Intangible Cultural Heritage. As Laurajane Smith points out, while Western monumental heritage is viewed as universally applicable and self-evidently having a universal audience, intangible heritage "tends to address much smaller audiences."[56] In the same vein, Barbara Kirshenblatt-Gimblett stresses how the exclusiveness of UNESCO's intangible heritage program, by "admitting cultural forms associated with royal courts and state-sponsored temples, as long as they are not European," preserves the majority view on heritage.[57]

While the initiatives discussed above regarding intangible cultural heritage took place outside the direct realms of the World Heritage Convention, UNESCO's general concern for intangible heritage has nevertheless filtered into the World Heritage discourse and practice. It should be regarded as at least implicitly stated in all the three developments already described: the Global Strategy urging development of a more anthropological view towards cultural heritage; the cultural landscapes category emphasizing the linkage between tangible and intangible values; and the Nara Document on Authenticity with its inclusion of traditions, spirit and feeling among the sources of information validating the authenticity of a heritage resource.

There has been a powerful incentive to transform World Heritage into a system focusing on the conservation of sites with both tangible and intangible values. It has been proclaimed that the outstanding universal value of a heritage place may also have immaterial manifestations and that these aspects should also be acknowledged in the World Heritage inscription process. Implementing this new perspective, however, has not been straightforward within the established structures of World Heritage. For example, a World Heritage Centre report, *Challenges for the Millennium*, when distinguishing between the 2003 Convention and the World Heritage Convention, asserts that the former's aim "to safeguard heritage that is living, in constant evolution and human-born" differs from "protection measures to tangible heritage, which often aim at preserving a specific state of conservation of a site."[58] Similarly, in 2004 the World Heritage Committee pointed out an essential difference between the World Heritage Convention and the 2003 Convention in relation to the concept of authenticity: "While World Heritage properties must fulfil the test of authenticity (...), intangible cultural heritage as defined by the 2003 Convention *is evolving continuously* and therefore a reference to the concept of authenticity was omitted."[59] In both these statements, the World Heritage Convention and the protection of tangible heritage were strongly associated with the notion of stability.

Enhancing the role of communities

From the mid-1990s onward there has also been a reformulation of policy in connection with the question of local community involvement in the definition and management of World Heritage. Prior to this date the relationship with local communities could best be described as elusive. In 1981 when an Australian non-governmental organization requested permission to provide additional material concerning the nomination of Kakadu National Park, the Committee decided that such groups would not be authorized to address it directly but instead only through national delegations.[60] In the 1992 version of the Operational Guidelines the States Parties were asked to refrain from "giving undue publicity to the fact that a property has been nominated for inscription" so "as to maintain the objectivity of the evaluation process and to avoid possible embarrassment to those concerned."[61] A reference to the "essential" local community participation during the nomination process was first introduced in the 1994 version of the Operational Guidelines.[62] After 1996, all references to the conditions concerning public participation were omitted, and the Operational Guidelines encouraged participation of local people in the nomination process in order "to make them feel a shared responsibility with the State Party in the maintenance of the site."[63]

According to Bart van der Aa, a key reason for the introduction of local pre-nomination consultation was to avoid potential local opposition after the designation, as, by the early 1990s, there had been cases of controversy over nominations in which the local population had been ignored.[64] This notwithstanding, the increasing role of local communities in the processes of World Heritage definition should not be seen merely as a result of a top-down design but equally as the reflection of the broader trend of 'democratization of heritage' and of people wishing to participate in developing a sense of cultural identity through heritage.[65]

Enhancing the role of communities was added as a fifth strategic objective of the World Heritage organization in 2007, based on the notion that community was key to the other four "strategic Cs,"[66] defined in the Budapest Declaration of 2002 as credibility, conservation, capacity-building and communication.[67] Three issues should be pointed out: firstly, the Community-C objective was included in the group of strategic objectives only as late as 2007. Secondly, when proposing the fifth "C," the expert group did not take a stand as to which level of participation would be proper and desirable with regard to World Heritage places, even though it distinguished between these different levels.[68] Thirdly, one cannot entirely escape the thought that in the discussion about the fifth "C" local community involvement was made discursively secondary to other objectives:

> Experience has demonstrated that one of the most important factors for the long-term success of a protected area, is *having the buy-in of affected indigenous/traditional and/or local populations*. […] Conversely, such communities which are disenfranchised may actively work against protected areas which do not reflect their interests, or *fail to deliver on the promises and/or expectations* raised when the site was given protected status.[69]

Similarly, in the Operational Guidelines local communities are made discursively secondary to other stakeholders. The document introduces World Heritage "parties" in a hierarchical order starting from the States Parties and moving to the

World Heritage Committee, the World Heritage Centre and the advisory bodies. Last on the list are mentioned the "partners" in the protection of World Heritage, who by definition "can be those individuals and other stakeholders, especially local communities, governmental, nongovernmental and private organizations and owners who have an interest and involvement in the conservation and management of a World Heritage property."[70] From the order of appearance and weight given, it is not difficult to see which are considered key stakeholders. It might be interesting to reverse this order and begin with a notion implying that communities are the main custodians of World Heritage. Also, the role of local communities is made discursively secondary to the official World Heritage parties by referring to partners and by using a low degree of modality (these groups "can be" partners).

Urban heritage as part of the definition of a "new era"

Through several interrelated stages, and in response to pressures from both inside and outside, the World Heritage community, from the early 1990s onward, renewed its philosophical basis in relation to what constitutes cultural heritage. This reassessment in the heritage paradigm may be understood as a movement from monumental and tangible perception of cultural heritage towards more anthropological, vernacular and intangible understanding. It may also be understood as a changed emphasis from the conservation of lost cultural traditions towards protection of living cultures. Moreover, the shift may be described as a partial reconsideration of a modernist definition of heritage on which the World Heritage Convention has essentially been based, and as a way of accommodating some of the concerns of its postmodernist critics. Finally, the early 1990s reassessment should be seen as a move towards decentralization of the system and acceptance of the more relativist idea of representativeness aside from those of outstanding universal value, universalism and globally applicable standards.[71] As has been pointed out on several occasions in this chapter, the proportion of this change is nevertheless open to question. Despite recent regionalization initiatives, the drawing up of the World Heritage List still resembles an endeavor to write an all-encompassing world history and an attempt to create a representative narrative in relation to human cultures.[72]

To open this discussion up to consideration of cities, let me now briefly touch upon the definition of a "new era" in relation to urban heritage at the level of the overall World Heritage policy. It first needs to be noted that cities were not at the heart of these concerns. This is not to claim that the new ways of thinking were in no way influential in the treatment of urban World Heritage in the immediate post-Global Strategy period; they surely were, as will be seen in the following chapters. Nevertheless, the focus of the re-articulation was elsewhere. This is not surprising, since historic cities, in the framework of the Global Strategy, were deemed to be an over-represented and somewhat outdated category of cultural heritage. The focus of interest for the identification of future World Heritage was placed elsewhere, on traditional and indigenous communities, on rural landscapes and on the linkages between culture and nature, because these aspects were identified as under-represented elements. Many of the novel conceptualizations of the early 1990s crystallized in the new nomination category of cultural landscapes, which was seen as being most effectively able to respond to the regional imbalances of the whole List. Although the three categories of cultural landscapes should be seen as conceptual rather than functional categories,[73] the World Heritage Committee's and ICOMOS' use of the concept has shown prioritizing towards a type of cultural landscape that was essentially rural or indigenous.[74] For instance, the expert group that discussed potential European World Heritage cultural landscapes in the mid-1990s did not mention cities even once in its final report.[75] As such, the project of cultural landscapes has been an anti-urban project. At the same time, in actual practice, between 1992 and 2002, the most common characteristic of a World Heritage cultural landscape was the presence of towns and villages.[76] It is not difficult to see how cities represent good examples of organically evolved continuing landscapes: they obviously result from an "initial social, economic, administrative, and/or religious imperative"; they have "developed their present form by association with and in response to their natural environment"; and they also "retain an active social role in contemporary society."[77] What perhaps makes the Operational Guidelines' definition of cultural landscapes less appropriate for cities is its reference to a close association with a traditional way of life. The question of how a traditional way of life can be maintained within rural or indigenous contexts is debatable,[78] but it seems especially difficult to associate with urban life.

The Global Strategy notion that the World Heritage List failed to reflect "living cultures" in part suggested that cities were to be primarily considered as architectural ensembles and not as representatives of "living cultures." What was also left aside at this stage was any reflection on how the old category of urban heritage could be associated with new meanings. A similar undertone is also detectable when reading the proceedings of the Nara Conference. While many of Nara's considerations dealt with the overall definitions of authenticity, and were thus, at least in theory, applicable to all forms of cultural heritage, there was very little specific reflection on authenticity in relation to cities.[79] The early stage formulations of authenticity in connection to cities remained influential. Michel Parent, for instance, had noted in 1983 that authenticity "cannot be judged solely in terms of the sum of the authenticity of the buildings that make it [the city] up. In a living town, the interpretation of modern life with the architectural heritage is such that this modern life can just as easily destroy the town's authenticity as enhance it."[80] This statement ambiguously suggested that there was more to the authenticity of a city than the sum of its buildings that is urban life, but that this modern life was detrimental to that very authenticity.

Concern for World Heritage cities nevertheless began to grow during the 1990s.[81] The main difference between the initiatives discussed earlier in this chapter and this growing concern for urban heritage was that while the first aimed at a conceptual reassessment, the latter was a rather reactive concern, motivated by concrete conservation threats. Since the end of the 1990s the World Heritage Committee had to deal with an accelerating number of large-scale development projects – especially high-rise construction and infrastructure and traffic projects – taking place within the boundaries of urban World Heritage sites or on their buffer zones. These cities included St. Petersburg, Istanbul, Cairo, Vienna, Santo Domingo, Luang Prabang, Edinburgh, Vilnius, Riga, Graz, Timbuktu, Liverpool and Dresden, to mention only a few.[82] The controversy involving the Historic Centre of Vienna, inscribed on the World Heritage List in 2001, and the planned Wien-Mitte urban development project located in an area that had been defined as a World Heritage buffer zone, ended up being influential. After having stressed on several occasions that if Austria would not change the course of the planned high-rise project, the Historic Centre of Vienna would be deleted from the World Heritage List, the World Heritage Committee, together with many local, national and transnational groups, was able to influence local planning in the way that the project was realized as a "light version," lower built and visually more compatible with Vienna's World Heritage status.[83]

At its twenty-seventh session in 2003, after the Vienna debate, the World Heritage Committee called for an expert meeting to discuss heritage conservation and socio-economic development in historic cities. The outcome of this meeting is called the Vienna Memorandum, a document that provides guidelines for both contemporary architecture and conservation management in the context of what the document calls "historic urban landscapes" versus the previously used term "historic centres." With regard to defining the relationship towards change in the context of urban World Heritage sites the Vienna Memorandum was a novel opening. For example, it is stated in the Memorandum that "continuous changes in functional use, social structure, political context and economic development that manifest themselves in the form of structural interventions in the inherited historic urban landscape may be acknowledged as part of the city's tradition."[84] Some commentators criticized this aspect of the Memorandum for providing a tool for actually legitimizing large-scale interventions for the sake of accepting change as part of historic urban landscapes.[85] However, when otherwise compared to earlier international recommendations in the field of urban conservation,[86] the Vienna Memorandum did not mean any revolutionary change of discourse. The definition of the historic urban landscape in the Memorandum was still much concentrated on the physical and visual city.[87] Similarly, the Recommendation on the Historic Urban Landscape, adopted by the UNESCO General Conference on 10 November 2011, is balanced between old and new ideas,[88] even though during its drafting rather open definitions of a historic urban landscape as an ongoing associative landscape, as a mindset and local knowledge and as "values that people bring into the city," were voiced.[89] These latest developments propose that urban heritage is in the process of catching up with the spirit of the moment.

Notes

1   UNESCO, World Heritage Committee, Thirteenth Session (Paris, 10–15 December 1989), Report of the World Heritage Committee. SC-89/CONF.004/12, Paris, 22 December 1989, p. 10, http://whc.unesco.org/archive/1989/sc-89-conf004-12e.pdf.

2 UNESCO, World Heritage Committee, Eighth Session (Buenos Aires, Argentina, 29 October – 2 November 1984), Report of the Rapporteur. SC/84/CONF.004/9, Buenos Aires, 2 November 1984, 3–6, http://whc.unesco.org/archive/1984/sc-84-conf004-9e. pdf; Operational Guidelines, January 1987, Paragraphs 23–31.

3 UNESCO, World Heritage Committee, Fifteenth Session (Carthage, 9–13 December 1991), Report, SC.91/CONF.002/15, 12 December 1991, http://whc.unesco.org/ archive/1991/sc-91-conf002-15e.pdf; Cleere, "The concept," 230.

4 UNESCO, World Heritage Committee, First Session (Paris, 27 June – 1 July 1977), Final Report, CC-77/CONF.001/9, Paris, 17 October 1977, p. 3–4, http://whc.unesco. org/archive/1977/cc-77-conf001-9_en.pdf; Cameron and Rössler, *Many Voices*, 45–82.

5 UNESCO, World Heritage Committee, Eighteenth Session (Phuket, Thailand, 12–17 December 1994), Report of the Expert Meeting on the "Global Strategy" and Thematic Studies for a Representative World Heritage List (UNESCO Headquarters, 20–22 June 1994), WHC.94/CONF.003/INF.06, Paris, 13 October 1994, p. 3–4, 6, http://whc.une-sco.org/archive/1994/whc-94-conf003-inf6e.pdf; concerning regional harmonization of national nominations, see Tanja Vahtikari, "From National to World Heritage via the Regional: Harmonizing Heritage in the Nordic Countries," in *UNESCO and World Heritage: National Contexts, International Dynamics*, ed. Casper Andersen and Irena Kozymka. Routledge (forthcoming).

6 Waterton, Smith and Campbell, "The Utility of Discourse Analysis."

7 Jokilehto et al., *The World Heritage List: Filling the Gaps*, 48–71.

8 UNESCO, World Heritage Committee, Eighteenth Session (Phuket, Thailand, 12–17 December 1994), Report of the Expert Meeting on the "Global Strategy" and Thematic Studies for a Representative World Heritage List (UNESCO Headquarters, 20–22 June 1994), WHC.94/CONF.003/INF.06, Paris, 13 October 1994, p. 5, http://whc.unesco. org/archive/1994/whc-94-conf003-inf6e.pdf.

9 Julian Smith, "Marrying the old with the new in historic urban landscapes," in *Managing Historic Cities*, ed. Ron van Oers and Sachiko Haraguchi. World Heritage Papers 27 (Paris: World Heritage Centre/UNESCO, 2010), 48.

10 See, for example, UNESCO World Heritage Committee, Eighth Session (Buenos Aires, Argentina, 29 October – 2 November 1984), Report of the Rapporteur. SC/84/ CONF.004/9 (Buenos Aires, 2 November 1984), p. 7–8, http://whc.unesco.org/archive/ 1984/sc-84-conf004-9e.pdf.

11 Robert Layton and Sarah Titchen, "Uluru: an outstanding Australian aboriginal cultural landscape," in *Cultural Landscapes of Universal Value*, ed. Bernd von Droste, Harald Plachter and Mechtild Rössler (Jena: Fischer, 1995), 176.

12 UNESCO, World Heritage Committee, Sixteenth Session (Santa Fe, United States of America, 7–14 December 1992), Report of the Rapporteur. WHC-92/CONF.002/12, 14 December 1992, 55, http://whc.unesco.org/archive/1992/whc-92-conf002-12e.pdf.

13 Mechtild Rössler, "Linking nature and culture: World Heritage cultural landscapes," in *Cultural Landscapes: the Challenges of Conservation*. World Heritage Papers 7 (Paris: UNESCO / World Heritage Centre, 2003, 10.

14 Operational Guidelines, February 1994, Paragraphs 36–38.

15 Operational Guidelines, February 1994, Paragraph 39.

16 Smith, "Marrying," 46. See also Jukka Jokilehto, "Reflection on historic urban land-scapes as a tool for conservation," in *Managing Historic Cities*, ed. Ron van Oers and Sachiko Haraguchi. World Heritage Papers 27 (Paris: World Heritage Centre/UNESCO, 2010).

17 Peter Fowler, "World Heritage Cultural Landscapes, 1992–2002: a Review and Prospect," in *Cultural Landscapes: the Challenges of Conservation*. World Heritage Papers 7 (Paris: UNESCO / World Heritage Centre, 2003), 18.

18 Henry Cleere, "The evaluation of cultural landscapes: the role of ICOMOS," in *Cultural Landscapes of Universal Value*, ed. Bernd von Droste, Harald Plachter and Mechtild

Rössler (Jena: Fischer, 1995), 57–58. See also UNESCO World Heritage Committee, Ninth Session (Paris 2–6 December 1985), Item 8 of the Provisional Agenda: Elaboration of Guidelines for the Identification and Nomination of Mixed Cultural and Natural Properties and Rural Landscapes. SC/85/CONF.008/3 (19 November 1985), p. 5, http://whc.unesco.org/archive/1985/sc-85-conf008-3e.pdf.

19 Herb Stovel, "Effective use of authenticity and integrity as world heritage qualifying conditions." *City & Time* 2: 3 (2007): 21–36, www.ceci-br.org/novo/revista/docs2007/CT-2007-71.pdf.

20 Jukka Jokilehto, "Authenticity: a General Framework for the Concept," in *Nara Conference on Authenticity: Proceedings*, ed. Knut Einar Larsen (Paris, World Heritage Centre, 1995), 19, 32.

21 Nobuo Ito, "'Authenticity' inherent in cultural heritage in Asia and Japan," in *Nara Conference on Authenticity: Proceedings*, ed. Knut Einar Larsen (Paris, World Heritage Centre, 1995), 36; Jokilehto, "Authenticity," 18; Dean MacCannell, *The Tourist: a New Theory of the Leisure Class* (Berkeley: University of California Press, 1999).

22 See, for example, Knut Einar Larsen, "Preface," in *Nara Conference on Authenticity: Proceedings*, ed. Knut Einar Larsen (Paris: World Heritage Centre, 1995), xiii. Ito, "'Authenticity,'" 35, discusses how the meaning of authenticity does not translate into Asian languages.

23 UNESCO, World Heritage Committee, First Session (UNESCO, Paris, 27 June – 1 July 1977), Issues Arising in Connection with the Implementation of the World Heritage Convention, CC-77/CONF.001/4, Paris, 9 June 1977, p. 8, http://whc.unesco.org/archive/1977/cc-77-conf001-4e.pdf.

24 UNESCO, World Heritage Committee, First Session (Paris, 27 June – 1 July 1977), Final Report, CC-77/CONF.001/9, Paris, 17 October 1977, p. 5, http://whc.unesco.org/archive/1977/cc-77-conf001-9_en.pdf

25 Operational Guidelines, October 1980, Paragraph 18.

26 Operational Guidelines, 20 October 1977, Paragraph 9.

27 Bernd von Droste and Ulf Bertilsson, "Authenticity and World Heritage," in *Nara Conference on Authenticity: Proceedings*, ed. Knut Einar Larsen (Paris, World Heritage Centre, 1995)," 4; Tanja Vahtikari, "Historic cities, world heritage value and change," in *Touring the Past: Uses of History in Tourism*, edited by Auvo Kostiainen and Taina Syrjämaa (Savonlinna: Matkailualan verkostoyliopisto, 2008), 112.

28 ICOMOS, No 30, June 6, 1978 (Warsaw, Poland).

29 UNESCO, World Heritage Committee, Third Session (Cairo and Luxor, 22–26 October 1979), Report of the Rapporteur on the Third Session of the Bureau of the World Heritage Committee (Cairo, 21 October 1979), CC-79/CONF.003/12 Rev., Paris, 30 November 1979, http://whc.unesco.org/archive/1979/cc-79-conf003-12reve.pdf.

30 UNESCO, World Heritage Committee, Third Session (Luxor, Arab Republic of Egypt, 23–27 October 1979), Comparative Study of Nominations and Criteria for World Cultural Heritage. Report by Mr. M. Parent, Paris, 20 September 1979, CC-79/CONF.003/11, Annex, p. 20, http://whc.unesco.org/archive/1979/cc-79-conf003-11e.pdf.

31 UNESCO, Bureau of the World Heritage Committee, Fourth Session (Paris, 19–22 May 1980), Report of the Rapporteur, CC-80/CONF.017/4, Paris, 28 May 1980, p. 4, http://whc.unesco.org/archive/1980/cc-80-conf017-4e.pdf.

32 UNESCO, World Heritage Committee, Third Session (Luxor, Arab Republic of Egypt, 23–27 October 1979), Comparative Study of Nominations and Criteria for World Cultural Heritage. Report by Mr. M. Parent, Paris, 20 September 1979, CC-79/CONF.003/11, Annex, p. 24, http://whc.unesco.org/archive/1979/cc-79-conf003-11e.pdf.

33 Operational Guidelines, October 1980, Paragraph 18.

34 UNESCO, World Heritage Committee, Seventh Session (Florence, Italy, 5–9 December 1983), Speech by Mr. Michel Parent, Chairman of ICOMOS, during the Seventh Session of the Bureau of the World Heritage Committee (Paris, 27–30 June 1983),

SC/83/CONF.009/INF.2, Paris, 1 September 1983, p. 4–5, http://whc.unesco.org/archive/1983/sc-83-conf009-inf2e.pdf.
35 von Droste and Bertilsson, "Authenticity and World Heritage," 7.
36 Jokilehto, "Authenticity," 17; Cameron and Rössler, *Many Voices*, 240.
37 Labadi, "Representations," 70.
38 UNESCO, World Heritage Committee, Eighteenth Session (Phuket, Thailand, 12–17 December 1994), Information Note: Nara Document on Authenticity. Experts Meeting, 1–6 November 1994, WHC-94/CONF.003/INF.008, 21 November 1994, http://whc.unesco.org/archive/1994/whc-94-conf003-inf8e.pdf.
39 Interview, Henrik Lilius, the chairperson of the World Heritage Committee in 2001–2002, December 9, 2002.
40 Operational Guidelines, February 2005, Paragraphs 82–83.
41 Feilden and Jokilehto, *Management Guidelines*, 17; Sophia Labadi, "World Heritage, authenticity and post-authenticity. International and national perspectives," in *Heritage and Globalisation*, ed. Sophia Labadi and Colin Long (London: Routledge, 2010), 71–72.
42 Feilden and Jokilehto, *Management Guidelines*, 60; Vahtikari, "Historic cities," 145.
43 Smith, *Uses of Heritage*, 54.
44 Early concerns regarding intangible heritage by UNESCO included, for example, copyright issues, ethnographic documentation, and the preservation of folklore. For discussion see Kirshenblatt-Gimblett, Barbara, "Intangible heritage as metacultural production," *Museum International* 56: 1–2 (2004): 53–54; Smith, *Uses of Heritage*, 106–107.
45 UNESCO, Encouraging transmission of ICH: Living Human Treasures, accessed December 15, 2015, www.unesco.org/culture/ich/index.php?pg=00061.
46 UNESCO, The Proclamation of Masterpieces of the Oral and Intangible Heritage of Humanity, accessed December 15, 2015, www.unesco.org/bpi/intangible_heritage/backgrounde.htm.
47 Chiara Bortolotto, "Globalising intangible cultural heritage? Between international arenas and local appropriations," in *Heritage and Globalisation*, ed. Sophia Labadi and Colin Long (Routledge: London, 2010), 97–114.
48 Ibid., 106–107.
49 See Logan, "Globalizing Heritage," 57, for an argument that UNESCO has accommodated change from the 'periphery.'
50 Smith, *Uses of Heritage*, 108.
51 UNESCO, The Proclamation of Masterpieces of the Oral and Intangible Heritage of Humanity, accessed December 15, 2015, www.unesco.org/bpi/intangible_heritage/backgrounde.htm.
52 UNESCO, *Convention for the Safeguarding of Intangible Cultural Heritage* (Paris: UNESCO, October 17, 2003), accessed July 12, 2015, www.unesco.org/culture/ich/en/convention, Preamble.
53 Francesco Bandarin and Sophia Labadi, *World Heritage: Challenges for the Millennium* (Paris: World Heritage Centre, 2007), 74.
54 Bortolotto, "Globalising intangible," 98.
55 See, for example, UNESCO, World Heritage Committee, Seventh Extraordinary Session (Paris, France, 6–11 December 2004), Item 9 of the Provisional Agenda: Co-operation and Coordination between UNESCO Conventions Concerning Heritage, WHC-04/7 EXT.COM/9, Paris, 25 November 2004, p. 7–8, http://whc.unesco.org/archive/2004/whc04-7extcom-09e.pdf, justifying the existence of two separate conventions on the basis of the essential differences between tangible and intangible heritage.
56 Smith, *Uses of Heritage*, 109.
57 Kirshenblatt-Gimblett, "Intangible heritage," 57.
58 Bandarin and Labadi, *World Heritage*, 73.

59 UNESCO, World Heritage Committee, Seventh Extraordinary Session (Paris, France, 6–11 December 2004), Item 9 of the Provisional Agenda: Co-operation and Coordination between UNESCO Conventions Concerning Heritage, WHC-04/7 EXT.COM/9, Paris, 25 November 2004, p. 8, http://whc.unesco.org/archive/2004/whc04-7extcom-09e.pdf.

60 UNESCO, World Heritage Committee, Fifth Session (Sydney, 26–30 October 1981), Report of Rapporteur, CC-81/CONF.003/6, Paris, 5 January 1982, http://whc.unesco.org/archive/1981/cc-81-conf003-6_e.pdf.

61 Operational Guidelines, 27 March 1992, Paragraph 14.

62 Operational Guidelines, February 1994, Paragraph 14.

63 Operational Guidelines, February 1996, Paragraph 14.

64 van der Aa, *Preserving*, 84–85. For examples regarding local communities turning down a possible World Heritage status see Aa, Bart J. M. van der, Groote, Peter D. and Huigen, Paulus P. P., "World Heritage as NIMBY? The case of the Dutch part of the Wadden Sea," in *The Politics of World Heritage: Negotiating Tourism and Conservation*, ed. David Harrison and Michael Hitchcock (Clevedon: Channel view publications, 2005); Vahtikari, Tanja, "Miten Vanhasta Raumasta tuli maailmanperintökohde?" in *Kotina suojeltu talo: Arkea, elämää ja rakennussuojelua Suomessa ja Saksassa*, ed. Outi Tuomi-Nikula and Eeva Karhunen (Pori: Kulttuurituotannon ja maisemantutkimuksen laitos, 2007), 111.

65 Millar, "Stakeholders," 39.

66 UNESCO, World Heritage Committee, Thirty-First Session (23 June – 2 July 2007, Christchurch, New Zealand), Item 13 of the Provisional Agenda: Evaluation of the Results of the Implementation of the Committee's Strategic Objectives. Proposal for a 'Fifth C' to be Added to the Strategic Objectives, WHC-07/31.COM/13B, Paris, 23 May 2007, p. 2, http://whc.unesco.org/archive/2007/whc07-31com-13be.pdf.

67 UNESCO, World Heritage Committee, Twenty-Sixth Session (Budapest, Hungary, 24–29 June 2002), Item 9 of the Provisional Agenda: The Budapest Declaration on World Heritage, WHC-02/CONF.202/5, Paris, 6 May 2002, http://unesdoc.unesco.org/images/0012/001257/125796e.pdf.

68 UNESCO, World Heritage Committee, Thirty-First Session (23 June – 2 July 2007, Christchurch, New Zealand), Item 13 of the Provisional Agenda: Evaluation of the Results of the Implementation of the Committee's Strategic Objectives. Proposal for a 'Fifth C' to be Added to the Strategic Objectives, WHC-07/31.COM/13B, Paris, 23 May 2007, note 8, http://whc.unesco.org/archive/2007/whc07-31com-13be.pdf.

69 Ibid., p. 4, my emphasis.

70 Operational Guidelines, 8 July 2015, Paragraph 40.

71 On the various aspects of change see Åsa Nilsson Dahlström, *Negotiating Wilderness in a Cultural Landscape: Predators and Saami Reindeer Herding in the Laponian World Heritage Area* (Uppsala: Uppsala University, 2003), 230; Labadi, "A Review"; Vahtikari, "Miten Vanhasta Raumasta"; Bortolotto, "Globalising intangible," 98; Cameron and Rössler, *Many Voices*, 94–95; Vahtikari, "From National to World Heritage via the Regional" (forthcoming).

72 See also Atle Omland, "The ethics of the World Heritage concept," in *The Ethics of Archaeology: Philosophical Perspectives on Archaeological Practice*, ed. Chris Scarre and Geoffrey Scarre (Cambridge: Cambridge University Press, 2006), 251.

73 Fowler, "World Heritage Cultural Landscapes," 18.

74 See also Peter Fowler, *World Heritage Cultural Landscapes 1992–2002*. World Heritage Papers 6 (UNESCO World Heritage Centre: Paris, 2003), 57.

75 UNESCO, World Heritage Committee (Merida, Yucatan, Mexico, 2–7 December 1996), Report of the Expert Meeting on European Cultural Landscapes of Outstanding Universal Value (Vienna, Austria, 21 April, 1996), WHC-96/CONF. 201 / INF. 9, 25 September 1996, p. 3, http://whc.unesco.org/archive/1996/whc-96-conf201-inf9e.pdf.

76 Fowler, "World Heritage Cultural Landscapes," 21.

77 Operational Guidelines, February 1994, Paragraph 39.

78 Nilsson Dahlström, *Negotiating Wilderness*.

79 A notable exception is Marc Laenen, "Authenticity in relation to development," in *Nara Conference on Authenticity: Proceedings*, ed. Knut Einar Larsen (Paris: World Heritage Centre, 1995), 351–357.

80 UNESCO, World Heritage Committee, Seventh Session (Florence, Italy, 5–9 December 1983), Speech by Mr. Michel Parent, Chairman of ICOMOS, during the Seventh Session of the Bureau of the World Heritage Committee (Paris, 27–30 June 1983), SC/83/CONF.009/INF.2, Paris, 1 September 1983, p. 6, http://whc.unesco.org/archive/1983/sc-83-conf009-inf2e.pdf. See also Operational Guidelines, January 1987, Paragraphs 23–31.

81 In 1993 the Organization of World Heritage Cities was founded independent from the official World Heritage organization. Organization of World Heritage Cities, accessed December 15, 2015, www.ovpm.org/. For the adoption of the World Heritage Cities Programme, see, UNESCO, World Heritage Committee, Twenty-Fifth Session (Helsinki, Finland, 11–16 December 2001), Report, WHC-01/CONF.208/24, Paris, 8 February 2002, p. 63, http://whc.unesco.org/archive/2001/whc-01-conf208-24e.pdf.

82 ICOMOS identified an increasing trend, between the years 1994 and 2004, in development threats concerning World Heritage sites, particularly in Europe and North America. ICOMOS, *Threats to World Heritage Sites 1994–2004: An Analysis*, May 2005, accessed June 12, 2014, www.icomos.org/world_heritage/Analysis%20of%20Threats%201994-2004%20final.pdf.

83 UNESCO, World Heritage Committee, Twenty-Sixth Session (Budapest, Hungary, 24–29 June 2002), Decisions Adopted, WHC-02/CONF.202/25, Paris, 1 August 2002, p. 37–38, http://whc.unesco.org/archive/2002/whc-02-conf202-25e.pdf; ICOMOS Austria, "The Wien-Mitte Project as Threat to the World Heritage Site 'Historic Centre of Vienna,'" in *ICOMOS World Report 2002–2003 on Monuments and Sites in Danger*, accessed April 10, 2012, www.international.icomos.org/risk/2002/index.html.

84 UNESCO, Fifteenth General Assembly of States Parties to the Convention Concerning the Protection of the World Cultural and Natural Heritage (Paris, UNESCO, 10–11 October 2005, Item 7 of the Provisional Agenda: Adoption of a Declaration on the Conservation of Historic Urban Landscapes, WHC-05/15.GA/INF.7, Paris, 23 September 2005, http://whc.unesco.org/archive/2005/whc05-15ga-inf7e.pdf.

85 "From the Permanent Delegation of the Federal Republic of Germany for UNESCO to Francesco Bandarin, UNESCO-Recommendation on the Historic Urban Landscape, Paris, 17 December 2010," accessed 7 January 2015, whc.unesco.org/document/117605.

86 UNESCO, *Recommendation concerning Safeguarding and Contemporary Role of Historic Areas* (Paris UNESCO, 1976), accessed March 3, 2016, http://portal.unesco.org/en/ev.php-URL_ID=13133&URL_DO=DO_TOPIC&URL_SECTION=201.html, and ICOMOS, *Charter for the Conservation of Historic Towns and Urban Areas* (ICOMOS: Washington, 1987), accessed March 3, 2016, www.icomos.org/charters/towns_e.pdf.

87 See also Smith, "Marrying."

88 UNESCO, Recommendation on the Historic Urban Landscape (Paris: UNESCO, 2011), accessed March 3, 2016, http://portal.unesco.org/en/ev.php-URL_ID=48857&URL_DO=DO_TOPIC&URL_SECTION=201.html. A more detailed analysis of the preparation and contents of the Recommendation falls outside the scope of this research. For such accounts see Michael Turner, "UNESCO Recommendation on the Historic Urban Landscape," in *Understanding Heritage: Perspectives in Heritage Studies*, ed. Marie-Theres Albert, Roland Bernecker and Britta Rudolff (Berlin: De Gruyter, 2013), 77–87; Sophia Labadi and William Logan, eds., *Urban Heritage, Development and Sustainability: International Frameworks, National and Local Governance* (London/New York, Routledge/Taylor & Francis Group, 2016).

89 van Oers, Ron, "Managing cities and the historic urban landscape initiative – an introduction," in *Managing Historic Cities*, ed. Ron van Oers and Sachiko Haraguchi. World Heritage Papers 27 (Paris: World Heritage Centre/UNESCO, 2010), 14.

4   World Heritage cities

Which urban pasts? Whose urban histories?

One of the dominant themes in the field of critical heritage studies concerns the politics of representation, something which has been studied in variable contexts from museums and heritage sites to postcolonial contexts, and at different levels, from local and state to supranational agencies. Frequently these studies have revealed celebratory and selective readings of history that focus on the representation of winners' history at the expense of minority versions of the past.[1] At the same time, many present-day museums and heritage sites manifest a genuine effort to display multiple and overlapping narratives.[2] Building on these insights, this chapter explores the ICOMOS discourse on World Heritage cities from the viewpoint of constructing historical value and narrating alternative pasts within the context of World Heritage cities. While it is obvious that not all histories of a place can be included in a concise evaluation document, it nevertheless suffices to ask which and whose pasts have been narrated and, equally important, which have not. Knowing that 187 cities from different parts of the world representing various epochs have been included in the World Heritage List up until 2011, the overall assumption must be that the ICOMOS discourse represents a great wealth and variety both in terms of chronology and in terms of various aspects of urban history – economic, political, social and cultural. By no means is this wealth and diversity disputed here altogether. Nevertheless, 'traditional' is the best word to describe the interpretation of urban history within the framework of ICOMOS evaluation texts.

Chronologies of urban World Heritage

Traditionalist attitudes versus "new towns of the twentieth century"

The early implementation of the World Heritage Convention was marked by a critical attitude towards modern architecture and urban planning. Brasilia, the modern capital of Brazil, inscribed on the World Heritage List in 1987 as a representative of the twentieth-century principles of urbanism, represents a unique case within the context of the urban submissions of the 1980s.[3] There had been an ongoing discussion regarding the identification of modern architecture in the context of World Heritage since the 1981 meeting of the Bureau of the World

Heritage Committee and the Australian nomination of the Sydney Opera House, for which unanimity could not be reached.[4] During subsequent years, the possibility of including modern architecture in the World Heritage List was not altogether rejected; however, the Bureau and the Committee on several occasions emphasized that a particularly strict evaluation process was necessary in relation to this category of heritage. It was feared that a general acceptance of contemporary architecture would quickly lead to an erosion and runaway expansion of the World Heritage List. Furthermore it was believed that an impartial evaluation of contemporary architectural structures was impossible due to the lack of a sufficient time distance.[5] Consequently, when formulating guidelines for the inclusion of historic towns on the World Heritage List in 1984, which later were contained in the 1987 version of the Operational Guidelines, the Committee made a restrictive statement concerning what it called the "new towns of the twentieth century": "History alone will tell which of them will best serve as examples of contemporary town planning. The files on these towns should be shelved until all the traditional historic towns, which represent the most vulnerable part of the human heritage, have been entered on the World Heritage List."[6]

This 'traditionalist attitude' was still common in the early 1990s,[7] and one of the conclusions made by the Global Strategy Expert Meeting in 1994 was that the twentieth century was an under-represented category in relation to other periods in the World Heritage List. The chronological-regional analysis conducted by ICOMOS in 2004–2005 showed that the World Heritage timeline was still similarly biased. In the analysis, note was taken for each site of the most significant period, or in certain cases two or three periods, mentioned in the justification of its outstanding universal value. Based on these divisions, the European Middle Ages received 152 mentions and the period from the Renaissance to the French Revolution another 90 in comparison to a total of 344 mentions for the entire European region (up to 1914), whereas the all-embracing universal category of "The Modern World" occurred only 14 times. The preference for pre-twentieth, and even pre-nineteenth-century heritage was clearly visible.[8]

If the same method of analysis were applied to the category of World Heritage cities, the results would be very similar. Thus from this perspective, a World Heritage city shows itself to be a predominantly pre-nineteenth-century city, when considered in terms of how the outstanding universal value has been justified. According to the World Heritage organization's own definition (put forward in July 2006) Valparaíso, Dresden, Liverpool and Cienfuegos should be regarded as modern nineteenth-century cities according to their primary reference, and Brasilia, Tel-Aviv and Le Havre as distinctively twentieth-century cities.[9] The Berlin Modernism Housing Estates and La Chaux-de-Fonds/Le Locle, inscribed in 2008 and 2009 respectively, should be added to that list.

When looking at the above-mentioned cities and the years of their inscription, a growing tendency in the new millennium to inscribe cities which were considered as also having had prominent roles in the more recent past may be discerned. Still, to a certain extent the question concerning the "new towns of the twentieth century" remains open as part of the official World Heritage policy: they have not made any rush to get on the World Heritage List, even though the later inscriptions of Tel Aviv, Le Havre or the Berlin Modernism Housing Estates did not arouse debates similar to those which once took place with respect to Brasilia. For example, the nomination of the Berlin Modernism Housing Estates was welcomed by ICOMOS as constituting "a valuable contribution to the World Heritage List, since modern housing estates are clearly under-represented so far."[10] Still, even the most recent Operational Guidelines, in congruence with the definitions created in the mid-1980s, retain twentieth-century cities in a secondary position.[11] It is indicative that at the time of its negative evaluation concerning the Argentine city of La Plata, built in 1882 as a capital city for the Province of Buenos Aires, and nominated as a representative of a planned new capital based on the realization of an ideal urban scheme, ICOMOS again recalled the Guidelines' remark concerning the granting of inscription to new towns of the twentieth century only "under exceptional circumstances."[12]

Although without a doubt indicative, these kinds of chronological classifications can also be considered problematic, since they require the determination of one or two 'golden ages' of development in the life of a city. For cities, such as Siena, labeled by ICOMOS "the embodiment of a medieval city,"[13] this distinction may be relatively easy, if not entirely unambiguous to make; for many other World Heritage listed cities, however, it is much more difficult, and may even be misleading. Fixing historical cities to a certain, limited, stage of their historical development easily promotes this one period in a city's history over all the other periods, at the expense of understanding how the city has grown during its gradual emergence.[14] While emphasizing one ideal phase of development, the later, or earlier, phases may seem like intrusions. This practice may also – in an unwelcome way – unify cities which share the same ideal phase but which during their later development have each moved, culturally, socially, economically or architecturally, in separate directions.[15] To prove the point concerning the somewhat arbitrary nature of chronological categorization applied by ICOMOS let me take Old Rauma as an example. Old Rauma was placed in the chronological-regional framework of the Gaps-report under the group "Europe / 15th and 16th centuries (Renaissance; Religious discords / Reformation; European colonization) / Northern Europe."[16] This is a very detailed categorization and Old Rauma alone represents this particular combination. This periodization, however, is strongly affected by the idea of an ideal phase of development. The reference to only the fifteenth and sixteenth centuries, with the fifteenth century marking the gradual founding of the city, fails to acknowledge the continuous urban development and the other significant periods of construction, urban life and historical identity. In Old Rauma's case a particularly important period of construction and formation of historical identity was the late nineteenth century, the so-called sailing-ship era and an era of urban prosperity, which involved active building in the Neo-Renaissance style. The existing building stock in the World Heritage area mostly dates back to this period.

At the same time, strict classifications contribute to the identified chronological under-representation. When viewed in less rigid terms, one is able to distinguish a broader group of cities within the World Heritage List than those identified above, which could be regarded, by slightly changing the perspective, as nineteenth-, and even twentieth-century cities. These cities include, for example, Nessebar, Bern, Saint-Louis, Budapest, Paris, Rauma, Sintra, Naples, Riga, Carcassonne, Guimares, Zanzibar, Baku, Vienna, Lamu, Essaouira, Acre and Bordeaux. This list is by no means exhaustive but it serves nevertheless to illustrate the difficulty of definition when stepping outside the obvious founding period in a city's history. Even Warsaw and Mostar could be considered twentieth-century cities within the framework of the World Heritage List, due to their almost complete post-war reconstruction. As noted by ICOMOS at the time of the nomination of Mostar, "what is visible will be substantially a product of the decades around 2000 AD."[17] Based on the above considerations it may be argued that some of the chronological biases identified with regard to urban World Heritage could be 'corrected' by reframing the narratives and chronologies of the existing World Heritage cities.

Cities in the time line

Moving beyond ICOMOS' chronological categorization of the World Heritage List to the question of representation of cities in relation to their own historical time line in ICOMOS evaluations, a similar ambiguous relationship regarding the more recent pasts becomes evident. On one hand, the more recent past and present existence of the inscribed urban area was often disregarded altogether. Here, too, the nineteenth and twentieth centuries were rarely seen as an integral part of the World Heritage city. On the other hand, when taken into account, the recent past and present were considered mainly problematic features.

Some early-period city inscriptions in particular promoted the idea of an end of history at a certain point in the past. For Valletta that point was the year 1798, after which "no important modifications," according to ICOMOS, had been made.[18] Moreover, it was rather common to apply a similar single-period approach, as described in relation to the Gaps-report's classification, disregarding the fact that all cities have been continuously inhabited and developing, and continue to be so.[19] This attachment of the outstanding universal value of cities to one ideal phase was, prior to 1994, supported by the cultural heritage criterion iv, which was often applied in the inscription of cities (see Appendix 3). The changes made in 1994 allowed for more flexibility in interpretation. Also, in those cases where historical stratification and chronological continuity were accepted as part of the definition of the outstanding universal value of cities in the early years, the scope of this approach remained limited. On these occasions continuity was interpreted mainly in reference to two arguments: firstly, when there were notable monumental architectural and archaeological ensembles of different styles dating from different periods within the inscribed urban area,[20] and secondly, when the historical continuity could be treated as a phenomenon of a far-enough-distant past. Consequently, Florence, valued as a "continuous creation over more than six centuries," was described only in relation to examples of building and architecture from the fourteenth to the seventeenth centuries leaving the rest of the "continuous creation" unmentioned.[21]

The successive stages of development and continuity leading up to the present were more often articulated in the context of value statements submitted since the late 1990s. At the same time, it may be argued that some of the earlier ambiguity remained. The more recent descriptions of urban pasts, despite hoping to articulate continuity, also often ended up highlighting earlier periods. For example, while ICOMOS noted the progressive and uninterrupted development of Verona through the incorporation of artistic elements from each succeeding period over two thousand years, the manner in which this continuity was described did not significantly differ from the earlier depiction of Florence. When introducing the most representative buildings, four periods in the construction of the city were identified: "the Roman period"; the "Romanesque period (eighth–twelfth centuries)"; the "Scaliger period (thirteenth and fourteenth centuries)"; and the "Renaissance to Modern period." The description of the latter, however, then focused almost solely on the fifteenth and sixteenth centuries with very little reference to succeeding centuries, except for a mention – itself quite rare in ICOMOS evaluations – of constructions by modernist architects Carlo Scarpa and Luigi Nervi.[22]

When taking a closer look into how ICOMOS treated the twentieth century as part of the description of history of the eight urban nominations inscribed on the World Heritage List between the years 2002 and 2003,[23] only two evaluation texts, namely those concerned with Valparaíso and Stralsund and Wismar, stand out as the kind which consider all the different historical periods as part of the relevant history. The former was discussed in light of its fortunes and misfortunes undergone from the beginning of the twentieth century to the world economic crisis of 1929,[24] and the latter in relation to early twentieth century population growth, World War II air raids and economic development during the German Democratic Republic period.[25] In the case of Paramaribo the twentieth-century past and the present were interwoven together in one sentence, which discussed the immigration of Indians and Javanese to Suriname as contributing to an increase in its cultural and ethnic diversity and as being "reflected in the present-day appearance" of the city.[26] It is notable how even the description of the history of Tel Aviv left the entire period after the 1930s, the iconic phase in the construction of the White City, virtually devoid of any discussion.[27]

Therefore, the general conclusion must be that the outstanding universal value of cities during the whole period of investigation was constructed largely as an antithesis to the recent and the current city. An especially difficult challenge has been posed by the post-World War II constructions. When built, they often marked an almost total change in relation to the existing urban structure. Today they, too, constitute an element of historical stratification – whether one likes it or not. The most permissive of ICOMOS evaluation texts towards the everyday heritage of the 1960s and 1970s within the World Heritage area is perhaps the account that ICOMOS issued with regard to Corfu in 2007, which stated that "these interventions represent a particular juncture in history and express the aesthetic attitudes of their time, clearly distinguished from previous buildings."[28] This layer, however, has most often been denied existence as part of the narrative of a World Heritage city.

Interestingly, contemporary architecture has, for the most part, been systematically treated by ICOMOS as belonging to the conservation history of the nominated area, and as something whose role is fulfilled only if it visually blends in with and is subordinate to the earlier forms and the existing appearance.[29] Overall, the framework within which ICOMOS has most often discussed twentieth-century history is as "Conservation history," which forms its own separate section in ICOMOS evaluations. This discussion has focused mainly on three aspects: the effects of the World War II on the built city, the post-war reconstruction in relation to international scientific standards of restoration, and the moment of acknowledging the conservation potential of the area. While all important considerations, as a byproduct, the twentieth century, or in many cases even the nineteenth century has rarely been given a status marking meaningful life or meaningful development. Therefore, it is refreshing to find a reference, even a brief one, to the present wine production in the area surrounding Bordeaux,[30] or to a regeneration of cultural and economic life in Liverpool during the recent decades.[31] In most cases the relationship between the past and the present-future in ICOMOS evaluations has been marked by discontinuity rather than continuity,[32] the conservation history often marking, discursively but clearly not in reality, the end of a long period of development.

Transnational and national narratives of outstanding universal value

Crossing boundaries

When viewing the World Heritage List in its totality, what constitutes a World Heritage city is first and foremost a political capital, a religious center or a commercial or fortified port. Between the years 1978 and 2011, 34 contemporary capital cities were inscribed on the World Heritage List, in addition to which the List obviously hosts several former political and administrative centers, such as Salvador de Bahia, the historical capital of Brazil (1549–1763), or Naples, the capital of the kingdom of Naples and Sicily (1738–1860), along with many regional and provincial capitals, for example Oaxaca, Quebec and Aleppo, as well as the 'European capital,' Strasbourg.[33] Apart from the obvious historical significance of capital cities as part of broader urban networks and in the building of modern nation states, the eagerness of states to nominate capital cities may also be linked to their projects of national-identity construction through nominations for inclusion in the World Heritage List. The capital card has been played "deliberately" by states,[34] and it has often been played at the early stage of involvement in the Convention. This was the case, for example, with the Baltic States that joined the Convention in the early/mid-1990s; Estonia, Latvia and Lithuania all chose to nominate their capital cities, Tallinn, Riga and Vilnius respectively, prior to naming any other sites.

Another popular inscription category since the early days of the implementation of the Convention has been that of the port city. Out of the six urban inscriptions in 2007, for example, three – Bordeaux, Corfu and the twin nomination concerning Melaka and George Town – were included in the List as representatives of international or regionally important seaports. Even though ports could be regarded as a well-represented category in the context of the present World Heritage List, it has been pointed out that certain kinds of ports, such as fluvial ports in Latin America, still continue to be an under-represented group.[35] It is not difficult to see why ports have been such a popular nomination and inscription category: by definition ports are international and open to the world, which makes them easily appreciated within a World Heritage discourse that prefers positive transnational interaction and geographical connectivity. The recent scholarship on transnational history emphasizes relations that, by definition, supersede national sovereignty and boundaries: "the transfers, linkages, interactions, and temporal as well as spatial flows between two or more national processes, events, or sites."[36] World Heritage designations use Wallersteinian world-system models to highlight past contacts between cultures and peoples,[37] and when viewed through the lenses of cities, and port cities in particular, it becomes obvious that establishing this interchange has been one of the main narratives that ICOMOS has wanted to support. The interchange of human values, the focus of selection criterion ii, has been a natural one to establish in reference to ports and other cities and, consequently, has been a frequently used justification criterion for their inscription (see Appendix 3).

One illustrative example of ICOMOS' eagerness to take up the argument concerning the interchange of human values between continents, as part of its statement of outstanding universal value, is the three times nominated, twice rejected and finally accepted twin nomination of the Spanish Renaissance cities of Úbeda and Baeza. These two cites were proposed for World Heritage candidature for the first time in 1989. On that occasion ICOMOS was of the opinion that in comparison to the five major Spanish historical cities already inscribed on the World Heritage List (Segovia, Santiago de Compostela, Avila, Toledo and Salamanca), Úbeda and Baeza, regardless of their architectural value, were of second rank.[38] In 2000, ICOMOS examined a revised nomination put forward by Spain, this time limited to the principal Renaissance areas. Once again, however, ICOMOS denied the inscription, as it felt that Úbeda and Baeza were not comparable to the best Italian examples of urban Renaissance (Pienza, Urbino). ICOMOS thus expanded the original comparative framework from Spain to the broader Mediterranean region. Nonetheless, in 2003, after a third nomination attempt, the two cities were finally considered to represent outstanding universal value, as they were now seen as bearing witness to the introduction of Renaissance design concepts into the Spanish context, and because these concepts, via Úbeda and Baeza, had become influential in Latin America. According to ICOMOS, several experts were consulted on the issue.[39] The interchange of human values that was associated with Úbeda and Baeza was understood within the traditional Eurocentric perspective of a one-way influence from Europe to the rest of the world. Nevertheless, as early as 1986, ICOMOS had accepted the nomination of Bolivia's Potosi along with a statement of significance that highlighted the key influence of Potosi's silver on the European economy ever since the sixteenth century.[40] This decision thus subverted the traditional representation of Europe as the main origin of influences and the center of world history.[41]

In addition to its articulation of interchange value, the process of searching for and finding the outstanding universal value of Úbeda and Baeza is also illustrative of other important issues. It shows that there has been a rather fine line drawn between successful and unsuccessful World Heritage nominations, and that outstanding universal value of a once-rejected site may be established after a thematic re-formulation, even a minor one. It also becomes clear that different experts at different moments can determine the World Heritage qualities differently, and that two comparative analyses may point toward diverse conclusions. What the chosen comparative framework may be is the key concern.

Finally, Úbeda and Baeza represent the relatively rare case of an urban serial nomination in the World Heritage List. States Parties, in general, have been reluctant to undertake such nominations, let alone transboundary serial nominations. At the time of the inscription of Stralsund and Wismar onto the World Heritage List, ICOMOS progressively suggested that Germany should consider the possibility of making the nomination of Stralsund and Wismar a serial nomination together with Lübeck, which had been awarded World Heritage status earlier. The rationale behind this measure, as explained by ICOMOS, was that all three towns were leading centers in the Wendish region of the Hanseatic League in Northern Germany, "representing complementary aspects in terms of trading, production of goods, and the typology of constructions."[42] Although Germany itself never acted on the proposal, several existing World Heritage cities could nevertheless be joined together for serial nominations. This would not only help to 'reorganize' the World Heritage List but could also contribute, at least partially, to diversification in its representational image. Going transboundary in making serial nominations would add yet another layer to the articulation of the interchange of human values.

Criterion vi

The idea of global heritage assumes identifying and valuing developments that can be considered global in their focus and character.[43] Here it is particularly warranted to look in more detail at the use of criterion vi (Table 2.1) in the context of the inscription of World Heritage cities. It is this criterion that assumes a connection with events, traditions and ideas of outstanding universal significance. On a general note it is difficult to distinguish any clear policy behind the use of criterion vi by ICOMOS and the World Heritage Committee, other than that this criterion has been the least often applied criterion, used on 36 occasions, that is 20 percent of inscription of cities, and that criterion vi was used more frequently during the early period of the implementation of the Convention. Both these considerations are in conformity with the World Heritage Committee's attempts to restrict the use of criterion vi over the years. However, what is also notable is that the 2005 objective to make the use of criterion vi again more flexible was not significantly advanced in 2006–2011. In fact, between 2001 and 2011, criterion vi was used only three times. Mostar became the first city ever included uniquely on the basis of criterion vi. In the cases of Macao and Cidade Velha, Historic Center of Ribeira Grande, criterion vi was introduced together with other criteria.[44]

The above-noted information suggests that the recent more flexible interpretation of criterion vi has not provided an avenue for a significantly different understanding of what constitutes a World Heritage city. Moreover, the rare use of criterion vi in reference to cities between the years 2001 and 2011 points to the fact that associations with architectural history, and much less with other historical associations, have continued to be considered the core of outstanding universal value for cities. The picture, however, is somewhat more diversified. Historical value, other than architectural historical, may be and has frequently been cited in reference to other criteria, especially criteria iii, iv and v. In fact, even though there is a marked difference between criteria iv and vi in their wording – in the case of the former, material heritage should be seen to be an outstanding example while illustrating significant stage(s) in human history, whereas, in the case of the latter, it is those events, traditions and ideas associated with the place that should be considered to be of outstanding universal significance – at times these two criteria have been used in parallel ways. For example, while the associations of the mining towns of Guanajuato and Potosi with world economic history were made in reference to criterion vi,[45] the very similar associations in reference to Liverpool representing the early development of global trading involved only the use of criterion iv.[46] Amsterdam's role as an intellectual, artistic and cultural capital in Europe in the sixteenth and seventeenth centuries was associated with criterion i, instead of criterion vi.[47]

When considering what kind of events, ideas and beliefs (as formulated before 1994), or living traditions and artistic and literary works (since 1994) associated with the place have been considered to be of outstanding universal significance over the years in the context of cities, two broad themes in particular surface as favored by ICOMOS: world religions, primarily Christianity and Islam,[48] and the "enlargement of the world," either involving the meaning of maritime exploration and the discovery of the New World by Europeans, or in reference to the development of global commerce and economic globalization.[49] More sporadically mentioned important themes and events include, for example, the opening of Russia to the Western world and the Bolshevik Revolution,[50] the suppression of slavery,[51] the unsuccessful attempt by Simon Bolivar to establish a multinational continental congress for newly independent American nations,[52] and the reconstruction of Warsaw and Mostar.[53] In the late 1990s, a minor boom was experienced in associating World Heritage cities with major artistic and literary works and their authors – examples include Prague, Salzburg and Vienna, all connected with musical arts, and especially with Wolfgang Amadeus Mozart.[54] Alcalá Henares and Prague were associated with the writers Miguel de Cervantes and Franz Kafka respectively,[55] and Bruges with the development in painting of the Flemish Primitives such as Jan van Eyck and Hans Memling.[56] The connection with major artistic and literary works was first specified in selection criteria in 1994; the intensified use of this reference during the late 1990s along with its complete abandonment later on, even though still present in the Operational Guidelines, shows that the justification for outstanding universal value has not been foreign to fashion. Also notable is that the artistic works and persons of outstanding universal significance were never identified within the context of contributions from outside Europe.

Finally, on a few occasions a general and unspecified remark concerning a city's association with universal history and universal importance was made. This was the case with cities, such as Rome, Venice and Saint Petersburg, which were considered as having 'indisputable' outstanding universal value. In reference to Saint Petersburg (Leningrad), ICOMOS stated that its inclusion on the World Heritage List "is so obvious that any detailed justification seems superfluous."[57] The 'indisputable' World Heritage value could also be justified by a generous application of the cultural criteria; sometimes the inclusion of an urban site in the World Heritage List invoked five or even all six criteria, as in the case of Venice and its Lagoon in 1987.[58] These statements about superior World Heritage value point towards the existence of an invisible hierarchy among World Heritage sites. They also suggest that certain cities have reached such a point in the process of their canonization, even prior to their World Heritage designation, that they are exempted from virtually any form of questioning. This corresponds both to what Dean MacCannel calls "sight sacralization,"[59] and to the early practice by World Heritage of listing of iconic sites clearly meeting the level of the "best of the best."

Reproducing national narratives

Joan Maddern calls for World Heritage sites to "promote themselves as transnational rather than as national spaces of citizenship, and seek to include rather than *police* ethnically or racially situated 'knowledges' and perspectives within their ideological borders."[60] As we have seen, the transnational and global turns in historiography have left their mark on the latest urban nominations and evaluations. The focus on transnational and global histories may be regarded as a consciously chosen counter-narrative to the one highlighting national past and identity. Still, despite the efforts of UNESCO towards postnationalism, the States Parties, by submitting nomination dossiers, have mainly wished to construct a narrative of "continuity, uniformity and stability of the nation."[61] Consequently, the ICOMOS discourse on cities has been a continuous balancing act between global and transnational and national histories. When reading the ICOMOS evaluations, it becomes clear that in many cases ICOMOS too has been willing to include national narratives, as proposed by the states, in its definitions of outstanding universal value. ICOMOS has not systematically associated outstanding universal value with postnationalism. This continuing significance of national narratives has been expressed in several ways: by direct references to a city's significance in the building of a national identity; by stressing, as part of the descriptions of cities, monuments, political traditions, famous historic persons and important historic events considered to be national icons, as well as the city's national significance in relation to planning, conservation and architectural history; or by the willingness to include capital cities in the World Heritage List, and also within their parameters many symbolically charged national monuments. Many examples could be pointed out. Direct references to the site's significance in the building of national identity, or to struggles for state independence include, for instance, Guimarães, associated with the "establishment of Portuguese national identity and the Portuguese language" in the twelfth century,[62] and Arequipa, whose role in the history of the Peruvian Republic was regarded as having been, and as continuing to be "crucial," as "a centre of popular civic rebellions and demonstrations, as well as being the birthplace of many outstanding intellectual, political, and religious figures in the country."[63] The struggle for Mexican independence in the early nineteenth century played a central role in the description of the history and outstanding universal value of Morelia in 1991: "Two of the leading figures in the struggle were both priests: Miguel Hidalgo and Jose Maria Morelos. In honour of the latter, a native of Valladolid, the town's name was changed to Morelia in 1828."[64]

The nation, grand monuments, heroes and nationally charged historic events, flavored with the global theme of finding the New World, were joined together in the evaluation document concerning Sintra:

> In the late 15th century Sintra was closely associated with one of the greatest queens of Portugal, Leonor, widow of Joao II, the "Perfect Prince." However, it was under the patronage of Manuel I that the town became indissolubly linked with the crown: he caused the Royal Palace to be substantially enlarged and founded the Monastery of Nossa Senhora da Penha, from which he watched the return of Vasco de Gama from his historic voyage. Succeeding monarchs spent much time in the town, and legend has it that King Sebastião listened to Camões reading his great epic poem *Os Lusiadas* there.[65]

That the States Parties, by submitting nomination dossiers, wish to construct a nationalist narrative, and thereby exclude more global and transnational histories, is not very surprising knowing that the building up of national identities has always been founded on ideas of national history, heritage and landscape, and that these constructions have always involved simplifications.[66] Perhaps it is somewhat more surprising that ICOMOS has chosen to reproduce some of these narratives of national identity as part of its own discourse involving cities. What becomes discernible here again is the UNESCO decision-making structure based on state representation and support, as well as the close-knit relationship between the ICOMOS valuation process and the state nomination dossiers. It has not been possible for ICOMOS to fully depart from these documents and the States Parties' representation of their 'own' pasts. It also seems that ICOMOS and UNESCO have had difficulties in expressing global place meanings and representing a supra-national World Heritage identity. A similar kind of complexity is pointed out by Wendy Beck in relation to travel guidebooks and their presentation of World Heritage sites. Contrary to what might be generally expected, global narratives have been largely omitted from the portrayal of World Heritage sites in travel guidebooks where the most commonly expressed theme is the destination's national significance.[67] A further explanation for ICOMOS' willingness to refer to national narratives may be more practical and relates to the scarce resources at the organization's disposal when making evaluations. In the case of limited resources it has been a straightforward task to reproduce certain arguments presented in the national nomination documents. Moreover, stressing national icons also corresponds to the interpretation of the early World Heritage cities as monuments (see Chapter 5).

Finally, the position of ICOMOS reflects the idea that global heritage can and should be inclusive of national, as well as regional and local heritages. In this connection it is important to consider what kind of national narratives have been promoted as part of the narratives of global heritage. There have been two kinds of nationalisms that the World Heritage organization has had to deal with. On the one hand, from the point of view of ICOMOS and the World Heritage Committee, it has been generally acceptable to include the national dimension as part of the definition of outstanding universal value and the representation of the site only if this connection has been expressed in a sensitive, positive and symbolic manner, an attitude exemplified by the cases quoted above and the frequent inclusion of capital cities in the World Heritage List.[68] On the other hand, as pointed out by Léon Pressouyre, the former ICOMOS World Heritage coordinator, those nominations concerning "symbolically charged national monuments, such as the National Monument at San Jose, Costa Rica or the Warrior's Cemetery and Monuments of Freedom of Riga, Latvia were judged incompatible with the principles of universality proclaimed by the Convention."[69] These two examples represent militant nationalism and thus are easy to exclude from the parameters of World Heritage; however, in many other instances, and when dealing with cultural heritage sites associated with highly diversified meanings, the distinction between the two readings of national histories and heritages (whether or not symbolically charged) is not as easy to make. The negative statements issued by ICOMOS show that ICOMOS has not rejected a single urban nomination based on the straightforward argument that the nomination advanced a too nationally charged political theme. Two nominations that perhaps came the closest in these terms were the Lithuanian nomination of Trakai Historical National Park and the Turkish nomination of the Old City and Ramparts of Alanya. Lithuania built the case for Trakai's nomination on its association with famous historical figures, the political significance of the town as the fourteenth century capital of the Grand Duchy of Lithuania and the role of the city in the formation of Lithuanian national consciousness, even though it also raised the theme of cultural co-existence. According to ICOMOS, what held the nomination together were the political features and "the memory of the golden era of the Grand Duchy of Lithuania" – qualities which were not sufficient on their own to convey outstanding universal value. That the tangible qualities of Trakai were not outstanding – rather than the nationalist framing of the site – was nevertheless presented as the main reason for the negative statement by ICOMOS.[70] The nomination of Alanya, which Turkey associated with the idea of the city's multicultural past as a place of long-lasting co-habitation of Christians and Muslims, was rejected by ICOMOS, since "both the churches and the Greek quarter, among the traces of Alanya's multicultural past, lie in despair."[71]

It is interesting to note that whereas a well-chosen national narrative has justified several successful urban nominations, the notion that the nominated city is of "great national value," as in the case of Trakai,[72] has often been used as an antithesis to outstanding universal value by ICOMOS when discussing rejected nominations. This is of course a well-designed strategy to avoid embarrassing the states involved in the procedure of rejection; the message is that even while not meeting the exceptional characteristics required by the Convention, the sites proposed by national governments are not without any value. On the other hand, by highlighting the national significance of the rejected nominations ICOMOS actually contributes to strengthening the overall nationalist narrative within the discourses of World Heritage. Therefore, rather than framing lacking outstanding universal value in relation to national value and issuing a vague statement about the nomination not meeting the criteria, ICOMOS should offer more detailed and nuanced thematic accounts as part of its negative statements. Happily, in its more recent statements of rejection, since roughly the mid-2000s, ICOMOS has indeed specified more systematically the reasons for its negation of outstanding universal value in relation to the actual criteria.

As noted in Chapter 2, the Convention spoke primarily, and contrary to other UNESCO conventions and recommendations at the time, in terms of 'heritage,' and not of 'property.' Sarah Titchen connects this choice of terminology to discourses on national versus international claims regarding heritage when she asserts that it implied a focus not so much "on the rights of a political sovereign over his national 'property,'" but rather on "the sense of duty to preserve and protect a 'heritage' inherited from the past whose values transcend national boundaries."[73] The later implementation of the Convention, as also noted by Titchen, has, however, blurred the use of this terminology, as it has become customary to speak about "properties" when listing sites as World Heritage. ICOMOS, too, even in its most recent evaluations, continues to use this terminology. This practice could be seen as something signifying ownership by the world community;[74] however, it is more likely that it supports the opposite national ownership concerns in the framework of World Heritage.

Marginalization of the view of urban history

The inclusiveness of heritages of a lower scale, however, as shown by ICOMOS in relation to national heritages, has encompassed fewer urban and local histories. I therefore argue that the very balancing between national and global narratives has meant a marginalization of the view of urban history, as understood by its academic practitioners, as something that goes beyond the "political history of the state and its overseas excursions,"[75] and as something where cities themselves are considered important instead of "the historical events and tendencies that have been purely incidental to them."[76] In ICOMOS evaluation documents cities have been portrayed less as independent and proactive actors than as subject to actions from outside and as dependent on larger processes. This is striking, even though it should of course be acknowledged that in the course of history the level of the independence of cities has varied greatly in different times and between different cultures.[77] An illustrative example in this regard is Derbent, the history of which was summarized by ICOMOS in the following way:

> The Persians (Sasanians) conquered the site at the end of the 4th century CE. The current fortification and the town originate from the 6th century CE, when they were built as an important part of the Sasanian northern limes, the frontier against the nomadic people in the north. From this time and until the 19th century, Derbent remained an important military post. From the 7th century, it was ruled by the Arabs, taken over by the Mongols in the 13th century, and by the Timurids in the 14th century. The Persians took it back in the early 17th century (the Safavid ruler Shah Abbas, whose capital was in Isfahan). In the 18th century, the Persians and Russians fought over Derbent, and finally the Russian sovereignty was recognized by the Persian Shah in the early 19th century.[78]

Furthermore, the interest of academic urban history in social and cultural history[79] has been relatively little reflected on as part of the ICOMOS discourse, apart from the elite narratives and the elaboration on the history of architecture and art. Consequently, several under-represented themes may be identified. These include, for instance, urban commercial culture, provision of housing, popular culture, working class history and urban migration. In addition, the descriptions of cities as World Heritage include relatively little elaboration with regard to the history of municipal, social and cultural institutions (hospitals, schools, museums, etc.) in cities. There are a few exceptions, such as Bruges, presented through its social institutions developed from the fourteenth century onwards.[80]

The re-nomination of Provins provides an exceptional thematic orientation. It has been relatively common for states to re-nominate once-rejected sites (17 times out of the group of 41), and while there are some examples of double rejections (Plovdiv, Sarajevo and Gdańsk), it should be noted that on several occasions the re-nominations have been successful (Appendix 2). Provins was first nominated under the title "the medieval town of Provins" in 1997. ICOMOS, in its negative assessment in 1998 noted that even though Provins was a well preserved and well managed town, and undeniably significant in the European context, it nevertheless lacked "that exceptional quality required for inscription." A cursory comparative analysis was made with regard to Visby and Bruges, in which it was suggested that Provins did not differ significantly from these two cities, the first already included in the World Heritage List and the second awaiting acceptance on the Belgian tentative list.[81] Following this negative statement the government of France decided to withdraw the nomination. The new proposal, put forward a couple of years later, was entitled "Provins, town of medieval fairs." According to ICOMOS, this second nomination had been "revised completely." This seems somewhat of an overstatement, since the two nominations and their successive ICOMOS evaluations included many identical elements. Not just the second nomination but also the first discussed Provins' status as a town of medieval fairs. However, it is true that the State Party was able to significantly sharpen the focus of the nomination and, consequently, was able to convince ICOMOS of its importance, showing that 'old-fashioned' heritage can be re-labeled to better fit with new expectations. The economic, social and cultural urban institution of fairs provided an integrating theme that allowed for a successful change of focus. Another important aspect was that the medieval institution of fairs was in essence an international phenomenon connected to long-distance interchange of merchandise between Europe and the Orient, which led in turn to an important sharing of human values.[82]

Another little-elaborated theme in ICOMOS evaluations has been urban infrastructure and services. The provision of piped water and sewerage, street lighting, electricity, or public transportation have rarely been included as part of the narrative of World Heritage cities. In the case of São Luís, portrayed as the first town in its region in Brazil to install a tramway system, to set up a water and electricity company, to light its streets with gas, and to have a telephone system, the wish to introduce these elements might relate to the city having been the first in these areas.[83] Discussing urban infrastructure would not only diversify the image of the World Heritage city but also allow reflection regarding other senses than the visual: for instance, sensing light, darkness, smell or noise in the city.[84] There are two important exceptions to this silence with regard to urban services. The first is the building and rebuilding of the street network, the Viennese *Ringstrasse* as an obvious example. The second is the arrival of a railway to a city, which has usually meant a major transformation in terms of urban networks and often caused visible changes in the urban fabric.[85] This points to the critical attitude towards change in ICOMOS evaluations, and will be discussed in more detail in Chapter 5.

Historical narratives embedded in political and economic history

Under-represented cities

The often-represented categories of capital and port cities within the World Heritage List, and the marginalization of the view of urban history already point toward certain conclusions regarding the representation of history. These have been confirmed by an in-depth analysis of the ICOMOS evaluation texts. The narrative regarding World Heritage cities has been firmly anchored in political and economic history, and in the histories of power politics between empires and states. To a significant extent, it has been a narrative of major developments, great events and ruling elites.[86] It is important to note that this overall emphasis on political and economic history may be linked to most World Heritage cities, not only to those credited with representing a significant political or commercial past. While not denying Naples' past political and economic role, the description of the history of this city, to cite one example, is illustrative of the above-mentioned points, as well as of our earlier discussion concerning the stagnant ideal of the end of the history of a city having occurred at a certain fixed date. The description begins with the founding of Naples in 470 BC after the battle between Cumae and Syracuse and the Etruscan fleet, and it continues with the entrance of the city into the Roman, and, later, into the Byzantine orbit. What followed was the Norman domination of the city, begun in 1139/40, and Hohenstaufen, Angevin, Aragonese, French and Spanish rule. After the Treaty of Vienna in 1738, Naples became the capital of the autonomous kingdom of Naples and Sicily, and, after a short period of French rule during the Napoleonic period, part of the kingdom of the Two Sicilies. The history of the city, as described in the ICOMOS evaluation, ends with the arrival of Garibaldi's army in Naples in 1860.[87]

At the same time, when using strict criteria, only one nomination can be pointed to that was inscribed on the World Heritage List between 1978 and 2010 as a purposed representative of an industrial city, even though industrialization contributed so profoundly to urban development from the late eighteenth century on, first in Europe and the United States, and later globally. The two Swiss towns of La Chaux-de-Fonds and Le Locle were awarded World Heritage status because of their urban and architectural association with the watchmaking industry.[88] Liverpool and Valparaíso are the other two urban nominations that come closest in these terms, but even in their contexts the use of the label 'industrial city' was avoided. In Liverpool's inscription the main focus of historical value was on the histories of a commercial port city of international importance, global trading and cultural connections and immigration to America, which was also reflected in the labeling of the site as a "Maritime Mercantile City."[89] In the case of the "Seaport City of Valparaíso," as the title suggests, the main emphasis in the application of evaluation criteria was similarly placed on the port city function and on the phase of early globalization, even though the evaluation text indicated elsewhere that Valparaíso could be understood to be "an exceptional example of heritage left by the industrial age."[90]

The limited representation of industrial cities obviously concerns the post-1800 period and the archetypical cities of the Industrial Revolution, but also to a lesser degree earlier manufacturing centers.[91] An important exception is that of a 'mining town,' several of which, mostly predating the high industrial era, have been added to the list since the inscription of the Norwegian town of Røros in 1980.[92] A minor aspect concerning the limited representation of industrial cities may be explained by the very definition of a World Heritage city applied in this research, and the requirement that the living 'urban' element play a significant part in the nomination, which excludes the three small-model industrial settlements of Crespi d'Adda, New Lanark and Saltaire, the Zollverein Coal Mine Industrial Complex in Essen as well as The Sewell Mining Town, deserted since the end of the 1990s. Further reasons, however, should be looked for elsewhere, and three such reasons in particular should be pointed out. Firstly, there has been the same refusal by traditionalists of cultural heritage at play that was discussed earlier in relation to modern heritage in general. Secondly, within the official World Heritage framework, as part of the typological categorizations concerning the World Heritage List and the national tentative lists, the two types of cultural heritage, 'industrial heritage' and 'historic cities,' have been retained as clearly demarcated and separate categories. They have not been joined together, even though both of them have been linked with other heritage types. While industrial heritage has been pointed out as one of the under-represented categories within the World Heritage List, industrial cities have not.

Thirdly, states may have become somewhat more careful recently to nominate larger-scale urban sites with development potential, which old industrial urban areas usually possess, because (as pointed out in Chapter 2) of the many debates concerning urban heritage versus development in the context of World Heritage–designated cities. Similarly, ICOMOS has become strict in its approval of the final delimitation of World Heritage areas. This careful approach, by all stakeholders involved, has led, and may lead in the future, to delimitation of industrial city inscriptions so that as little as possible dissonance may occur, for example by being delimited with regard to the industrial and technological complexes and structures without the wider urban structure being involved.[93] Among the cities examined in this research the reluctance to include industrial pasts in the narrative of World Heritage cities through the means of physical delimitation of the World Heritage area can be seen, for example, in the case of Baku. Up to the present, Baku has been one of the major centers of oil production in the world. The area testifying to the late nineteenth and early twentieth century phases in this development within the city was not, however, included in the World Heritage zone. Instead it was decided that this area should form a buffer zone.[94]

For very few cities has their outstanding universal value been constructed in relation to their industrial past, other than through a passing reference. Congruent with the chronological under-representation of the more recent pasts, industrialization of cities was for a long period an almost completely silenced theme with respect to the discourse on World Heritage cities and their outstanding universal value. Or it was only treated as something that was a violent rupture in the life of the city. In this respect, some of the more recent ICOMOS evaluations have also begun to show more inclusive and neutral commentaries. Therefore, ICOMOS wanted to include in the World Heritage area of Riga "the impressive heritage of workers' housing" from the nineteenth and twentieth centuries as a witness to Riga's past as a major industrial city.[95] In the case of Dresden, considered to represent land use during the process of early industrialization in Central Europe, the "industrial heritage consisting of the remarkable steel bridge, the rare historic railways and the historic steamships" was considered to "complement the ensemble."[96] Still, these commentaries, as also exemplified by the cases referred to above regarding Liverpool and Valparaíso, remain modest, and secondary to other versions of the past. Without denying that the change in urban form and urban life caused by industrialization was abrupt in many cities, excluding these pasts almost entirely furthers a stereotypical view. Obviously the question of representing industrial pasts is not relevant concerning all World Heritage–designated cities, since some of them either were never significantly affected by industrialization, or have had a very clearly defined, earlier period of significance which has been identified as the basis of outstanding universal value. The same, however, seems to be true for cities the outstanding universal value of which was justified essentially in reference to their long-term continuity, as for example with Oporto or Naples, both having supported an important industry during the course of their history. It is in cases like those of Oporto and Naples, or Baku, where discussing the city's industrial pasts would, while perhaps not conforming with the public images of these cities, and while perhaps complicating the picture, also make statements concerning outstanding universal value significantly richer.

In tandem with the under-representation of industrial cities, the typical World Heritage city has so far failed to a large extent to be a 'recreational city' or a 'suburbia.' Recreational cities, such as seaside or skiing resorts, reflecting the development of tourism industries and displaying many distinctive features, such as seasonality of business, dominance of the service sector or the "architecture of pleasure,"[97] have thus far not been represented among World Heritage cities. As noted by Peter Clark, a few specialized leisure towns existed in Europe before 1800; the majority of such towns, however, emerged from the 1830s onwards. Consequently, by the end of the nineteenth century there were already hundreds of spa towns in Europe.[98] There have been some local- and national-level discussions about nominating Blackpool for World Heritage inscription as a representative of the "world's first working-class seaside resort,"[99] but until today it has not been added to the tentative list of the United Kingdom.[100] The history of Innsbruck as the site of two Olympic Winter Games, in 1964 and 1976, was not able to convince ICOMOS of the city's outstanding universal value, and the organization ended up rejecting the nomination in 2005, giving as its reason that the overall integrity of the Valley of the Inn River was compromised.[101] However, it should be acknowledged that there are currently plans for a transnational project to nominate a selected group of spa towns to the World Heritage List under the title "Great Spas of Europe," to complement Bath, which was inscribed in 1987.[102]

Overall, the history of tourism and travel, even though prominent in the context of many World Heritage cities, now as before, has rarely been portrayed. Two of the few exceptions in this respect are Assisi, highlighted as one of the major places for visitation in Italy from the Middle Ages up to the present,[103] and St George, discussed in the context of tourism in Bermuda.[104] Tourism and heritage are two interlinked phenomena; the tourist as a consumer contributes to the creation of heritage.[105] Even though UNESCO and ICOMOS have acknowledged this link, in particular since the early 1990s,[106] the most recent ICOMOS evaluations still continue to refer to tourism mainly within the framework of analyzing future risks to World Heritage sites under the title "tourism pressures." The tourist's role in creating place meanings does not make up part of the discussion.

The failure to thus far represent the history of suburban development has become equally evident. A World Heritage city has been understood almost solely to be a historic city center inclusive of early urban enlargements but exclusive of nineteenth- and twentieth-century suburban developments, thus again implying a discontinuity with regard to more recent phases. Only a very few exceptions to this prevailing city center view may be pointed out. The Jewish Quarter of Třebíč in the Czech Republic was inscribed, without including the rest of the historic city center of Třebíč, as a representative of an ethnically segregated area formed in the course of history, but obviously not to be considered as a representative of twentieth-century suburbanization.[107] Another exception to the prevailing city center approach to outstanding universal value is the Dresden Elbe Valley, treated as an urban cultural landscape, in the case of which the city center and parts of the adjacent suburbia were considered to form an entity.[108] This type of approach calls for a broad delineation of the World Heritage area. Paradoxically, it was Dresden that, partly because of the broad delineation of the World Heritage area, got deleted from the List in 2009, when the city decided to build a car bridge in the midst of the urban cultural landscape.[109] Finally, the Berlin Modern Housing Estates, which consists of six housing ensembles constructed between 1913 and 1934, most unambiguously represents suburban history among the 187 urban sites studied. It is unclear, however, whether Germany and ICOMOS consider this site distinctively urban; in the ICOMOS evaluation it was primarily discussed in relation to the developments of modern architecture, thus transmitting a rather monumental understanding of the site.[110] Nevertheless, I consider that this site at least has the potential to further a suburban interpretation of a type of World Heritage city at present still mostly absent from the World Heritage List.

It is hardly surprising, since the World Heritage organization has been founded with the objective of architectural conservation, that the ICOMOS historical narrative regarding cities has dealt extensively with the history of architecture, urban planning and construction. It is the area where ICOMOS' expertise lies and about which it is expected to elaborate. That architectural history lies at the heart of World Heritage value was made explicit with regard to Ouro Preto: "As first capital of the state of Minas, Ouro Preto is of local interest; as a mining centre of the Golden Age of Brazil, it is of national interest, and as a unique centre of baroque architecture, it is of outstanding universal value."[111] As ICOMOS evaluations have become more detailed over the years, so have the descriptions concerning architectural history. There have, however, been significant differences in how the relationship between the history construction and other histories has been emphasized and explained. On the one hand, there are examples such as Morelia, inscribed in 1991, for which ICOMOS' only commentary regarding the history of construction was a one-sentence note about the establishment of a Franciscan monastery. Otherwise the evaluation text focused on aspects of political and economic history.[112] On the other hand, in many cases, including some recent ones, the history of architecture and building were discussed in a manner almost completely detached from the historical development of the city in its other areas of life. One example is Arequipa, inscribed in 2005, for which references to themes other than the history of architecture were so scarce that it is somewhat difficult to find support for its ultimate inscription on the basis of criterion v as an "outstanding example of a colonial settlement, challenged by the natural conditions, the indigenous influences, the process of conquest and evangelization, as well as the spectacular nature of its setting."[113] In addition to Arequipa, the recent architectural-history-dominated evaluation texts include especially Yaroslavl, Úbeda and Baeza, and the late-baroque towns of the Val di Noto.[114] In reference to the latter, ICOMOS regretted that the nomination by the state had been "almost entirely based on 18th century urban art and architecture and says nothing about urban economy or urban/rural relationships"; the organization nevertheless chose to reproduce this narrative proposed by the state as part of its own evaluation document, which suggests ICOMOS' ultimate will to accept it. This example also illustrates the difficulties that ICOMOS faces when trying to depart from the state-sanctioned narratives. ICOMOS and the Bureau of the World Heritage Committee asked Italy to supplement its initial nomination of Noto with a management plan. No commentary, however, was made on this occasion with regard to the description of history in the nomination dossier.[115]

His-stories of ruling elites

Sophia Labadi argues that the World Heritage nomination dossiers by states, whether concerning religious or industrial heritage, are highly androcentric and un-democratic representations, as they put emphasis on the "great men of history," while at the same time marginalizing both men from lower classes and the historical and contemporary female. This stands in contrast to the rhetoric of several UN and UNESCO declarations, which highlight the need for sufficient recognition of women's contributions in the global arena. Labadi further notes that when mentioned in national nomination dossiers, women and the lower classes are described in stereotypical, superficial and depersonalized ways.[116] These findings are congruent with the notion that even though there exist plenty of guidelines and heritage-related literature that take gender issues seriously, gender continues to be overlooked in discussions of heritage. Places of significance to women's history are often absent from heritage registers, and the representation of heritage tends to legitimize gender stereotypes of men and women, and the social values that underpin them. Heritage "tells a predominantly male-centred story, promoting a masculine, and, in particular, an elite-Anglo-masculine, vision of the past and present."[117]

It is not difficult to find a similar male-based language and emphasis on upper- and middle-class narratives when viewed through the lenses of ICOMOS evaluations – the recognition of female agency by gender history and gender archaeology has not filtered into the World Heritage discourse. Even though a few evaluation texts, such as those concerning the mining towns of Zacatecas, Guanajuato and Potosi, did not mention important personalities at all,[118] most did, time and time again. The references made were primarily to kings and emperors, popes, founders of the city, governors, explorers, artists, architects and urban planners, thus, with very few exceptions, all men. As Dolores Hayden notes, urban heritage often revolves around architectural monuments associated with male elites as "city fathers."[119] While being no less upper class, Saint Petersburg probably represents the most 'female city' among World Heritage inscribed cities, as references were made to the prominent empresses of Russia, Anna, Elisabeth and Catherine.[120]

Even though the references to important personalities in ICOMOS evaluation texts often concern internationally renowned men, there are also examples such as Goiás, whose less-famous explorers and governors are meticulously identified.[121] Aside from the influential historical male figures, references to contemporary professionals, such as Michael Levin, the organizer of an international exhibition on

Israeli modern architecture, can be found.[122] In the case of Verona, the long list of rulers, architects and artists was closed with a reference to conservationist Piero Gazzola, who was also the founding President of ICOMOS.[123] Warsaw, Bruges and Mostar also provide similar examples of legitimizing the professional authority of conservationists in ICOMOS evaluations.[124] Several non-Western examples, such as Lamu, listing the town planners of the 1970s and 1980s,[125] show, in contrast to Labadi's findings,[126] that references to great men have not been limited to evaluations concerning Western cities only.

This focus on the history of the life of a hero and on upper- and middle-class male narratives as part of the construction of outstanding universal value of cities by ICOMOS seems surprising for two reasons. Firstly, because one would generally expect a less individual-focused approach in the context of cities and urban history in comparison to monumental heritage; and secondly, because (male) heroes figure prominently on the agenda of nationally focused history writing and heritage projects,[127] and would hence seem to fit less comfortably with the projects of transnational history promoted by ICOMOS and UNESCO. As noted in Chapter 2, in 1979 the World Heritage Committee already thought that the selection of sites connected with famous people could be "strongly influenced by nationalism."[128] It is noteworthy that gender has never been identified as an underrepresented theme within the framework of World Heritage, most likely because of "the objective of universal equality, which does not accommodate and regulate the treatment of difference."[129]

Nevertheless, what is most striking is that while there are obviously fewer references to persons in the more concise ICOMOS evaluations than in the much longer national nomination documents, ICOMOS seems to have omitted most of the references to women made by states. For example, Italy in its justification of outstanding universal value for Ferrara explained that "outstanding female personalities of the Este family such as Beatrice and Isabella d'Este or Lucrezia Borgia contributed to the prestige of the court." ICOMOS, however, did not include this aspect within its own analysis.[130] Around 30 percent of national nominations present "women in a more egalitarian manner as public figures."[131] A close reading of the 16 ICOMOS urban evaluations carried out from 2001–2003 reveals that women were mentioned hardly at all, let alone described in an egalitarian manner; that occurred only 3 times in comparison to 89 references to men. Either urban history is seen as a dominantly male field by all stakeholders involved, or ICOMOS, when summing up the core of the national nominations, leaves the role of women unmentioned. Both are unsettling images, contributing to the construction of women as the 'invisible other' within the framework of heritage. As pointed out by Laurajane Smith, this construction of gender identities in heritage "has a range of implications for how women and men and their social roles are perceived, valued and socially and historically justified."[132] Recently, however, the World Heritage Committee called for the use of gender-neutral language in association with the statements of outstanding universal value.[133] This suggests that gender is finally appearing on the World Heritage agenda.

Since the narrative on World Heritage cities has tended to focus on the history of ruling elites, it has also failed to be the history of ordinary urbanites, the working class or of urban migrants or other marginal groups in the city. We get a glimpse of these marginal groups, and the controversial histories they represent, in connection with Saint Petersburg, where a "colossal forced labour of Russian soldiers, Swedish and Ottoman prisoners of war, and Finnish and Estonian workers and laborers" contributed to building the city,[134] or with Potosi, where by the seventeenth century there were "160,000 colonists and 13,500 Indians, who were forced to labor in the mines" nearby.[135] In association with Lyons, a reference was made to the life of silk industry workers and to the first workers' demonstration.[136] However, most descriptions mention very little of this; the histories of rulers and urban elites is further supported by frequent references to noble palaces and villas, and, conversely, by many fewer references to ordinary burghers' houses. This emphasis on the history of the ruling elites is not entirely surprising even in light of the recent trends identified in urban historiography, which continue "to privilege the world of the bourgeoisie and the more traditional urban elites."[137]

Particularly important seems to have been the identification of the founder(s) of the city. This fits into one of the overall conclusions of this research that finds that the ICOMOS discourse on World Heritage cities has been very uniform and standardized. While each city is investigated in line with its own past, the developmental trajectories of cities are nevertheless presented in a very formalized way. The description of history usually begins with the first inhabitation of a region in small settlements. What often follows is a description of the official founding of the city, the first written records about it, and the establishment of the original town plan. In the case of colonial cities, an important phase is, of course, the discovery by Europeans. The description then moves through the cycles of political upheavals, economic growth and prosperity, to the hardships and disasters confronted by the city and the subsequent effects on building and architecture. The story often ends with the city's ultimate decline as an economic (and social) entity and the transformation of the place into a heritage resource. These are linear and evolutionary presentations of history, based on Western models and concepts.[138] In the process, the complex lives of cities are simplified and abstracted. One significant factor that contributes to this uniformity is of course the standardized process of nomination, evaluation and designation. Specific guidelines have been written to inform states of how to make nominations. The representatives of states, or the consultants who write the nominations, look for guidance from earlier successful proposals. All nominations are evaluated against uniform qualifying criteria. All these predestined elements produce uniform narratives with the potential to hide the localness, heterogeneity and particularity of individual places. Local peculiarities and values, or local discrepancies, do not sit easily with the universalizing definition of outstanding universal value. It can be argued, however, that a diametrically opposite approach, one emphasizing heterogeneity, would fit well with the objectives of heritage diversification outlined in the framework of the Global Strategy and the Nara Document on Authenticity.

Unanimous urban heritage

Uniform narratives also often hide the contested nature of individual places and produce an image of a harmoniously shared heritage of humankind. When referring to a past that is recognized as meaningful in the present, but that is also contested and awkward, Sharon Macdonald uses the concept "difficult heritage."[139] Lynn Meskell uses the term "negative heritage" in a similar manner to describe "a conflicting site that becomes the repository of negative memory in the collective imaginary."[140] There are only a few cultural sites on the World Heritage List that fall within this definition. These include, most importantly, the Auschwitz Birkenau concentration camp, the Island of Gorée, the Hiroshima Peace Memorial, Robben Island and the Bikini Atoll Nuclear Test Site in the Marshall Islands. It has been argued, however, that the actual inscription of these sites on the World Heritage List has been founded on other, less contested aspects.[141] Moreover, as Atle Omland points out, the war-related sites have usually been treated as peace memorials within the World Heritage framework.[142] UNESCO's peace message and the ideal of a shared heritage of humankind assume this kind of harmonious interpretation and educational agenda. The organization's reluctance to discuss difficult heritage should therefore be primarily understood as an attempt to avoid open politicization.[143] Trinidad Rico calls this "a delicate negotiation around negativity that is needed to subsume sites into a fixed set of criteria as well as a set of approved political issues."[144] She argues that the World Heritage List, to maintain its role as a credible educational archive, should include more heritage sites with negative associations.[145]

Not surprisingly, the associations with difficult heritage also remain few within the context of ICOMOS' portrayal of World Heritage cities. Only two cities – Warsaw and Mostar – can be identified as war-peace memorials, even though Dresden, for example, could also have been identified this way. In the case of Warsaw the war-peace association was made by referring to "the will of the nation" which brought to life again a city from the ruins.[146] However, the outstanding universal value of both these cities was defined by ICOMOS mainly in relation to aspects other than negative ones. The main focus of Warsaw's inscription was placed on the development of scientific restoration in the post-war period.[147] Mostar, first nominated for inclusion on the World Heritage List by Bosnia and Herzegovina in 1998 and added to the List in 2005, was similarly valued as a universal symbol of human solidarity for peace, reconciliation, international cooperation and the "coexistence of diverse cultural, ethnic and religious communities."[148]

Furthermore, it is interesting to note a case – that of Sarajevo – in which ICOMOS unequivocally refused to comment on a negative association proposed by the State Party. Despite the two subsequent nominations in 1985 and 1998 by Yugoslavia, ICOMOS never recommended Sarajevo's inclusion in the World Heritage List. In the 1985 nomination dossier, the Yugoslavian authorities argued the case for Bascarsija, the historic center of Sarajevo, on the grounds of the historical significance of the city as the symbolic setting for the outbreak of the World War I. ICOMOS' negative recommendation was based on the notion that there existed other, more representative examples of an Ottoman city in Europe. No reference was made by the organization to Sarajevo's historic association with World War I.[149] The second nomination, by Bosnia and Herzegovina in 1998, and made in the aftermath of the 1992–1995 war, shifted the focus to Sarajevo as "the guardian and the symbol of the multi-cultural way of life." ICOMOS recommended non-inscription on the grounds that Mostar, also nominated by Bosnia and Herzegovina in the same year, was a "superior" candidate for inscription, mostly because of its "spectacular location."[150]

Multiculturalism has regularly been an advocated theme for signaling outstanding universal value for cities – a natural interpretation, since cities, more than rural areas, may be considered as places for the coming together of different peoples, of heterogeneity and of intercultural dialogue. Multiculturalism in the ICOMOS discourse has most often denoted a historical phenomenon, either multiculturalism based on consecutive cultures in a city – Baku for instance was appreciated for revealing evidence of "Zoroastrian, Sassanian, Arabic, Persian, Ottoman, and Russian presence in cultural continuity"[151] – or the co-existence in the past of several ethnic and/or religious groups in the city.[152] In the case of the Slovakian town Bardejov, ICOMOS took the initiative to advance the idea of multicultural heritage in the local community by requiring, in contrast to the original nomination by the Slovakian state, that the World Heritage area should also include the city's small Jewish quarter.[153] The Czech town of Třebíč where the old dilapidated Jewish district and cemetery were, within a relatively short period of time, restored and revalorized as symbols of multiculturalism, can be seen as a case in which multiculturalism served as a successful inscription strategy.[154] The failure of the State Party to convince ICOMOS of the ongoing multiculturalism in urban society led to negative statements in the cases of Alanya and Jajce.[155] In its emphasis on multicultural heritage, ICOMOS has followed the trend of recent years concerning the growing recognition by states and other agents of the need to promote multicultural policies. The great potential seen in the interpretation of multicultural heritage is the ability of this heritage to produce positive values leading to greater social inclusion and harmony.[156] On the other hand, the "banner of multiculturalism" may be something that itself creates a problem, since it provides an unchallenging means for celebrating cultural diversity without "engaging with the realities of racist and unequal relationships between ex-colonial and ex-colonized countries, and between genders and classes."[157]

Within the World Heritage List Mostar represents a city for which multiculturalism in the present society, together with or even beyond its multiculturalism in the past, became the prime focus. Still, at the moment of Mostar's inscription, the proclaimed multiculturalism and reconciliation was far from being an unambiguous local reality, since the city was starkly divided along ethnic lines. In fact, the image of Mostar as consciously multicultural has been shown to be in large part a wartime and post-war construction projected backwards onto the pre-war period, and essentially founded on the imagination of international organizations and the media.[158] Even though the city was administratively united in early 2004, this took place by a decree of the international community's high representative, Lord

Paddy Ashdown, and not on the basis of an actual agreement between the city's two ruling parties – the nationalist Bosniak (Bosnian Muslim) Party for Democratic Action (SDA) and the Croatian Democratic Union (HDZ).[159] Also at odds with the notion of multicultural heritage in the context of the Mostar World Heritage site was its restrictive delimitation: the inscribed area covers only 7.2 hectares, consists predominantly of Ottoman-era (historical or reconstructed) heritage and is inhabited mainly by Bosnian Muslims. This delimitation reflected an essentially monocultural view, something which was nevertheless not addressed in the official evaluation documentation. Instead, this document elaborated extensively on the issue of the authenticity of Mostar after its reconstruction.[160] Paradoxically the notion of multicultural heritage based on the idealized version of the past did very little to include the Croat community.

Underlining multiculturalism in regard to Mostar may be considered part of ICOMOS' strategy to downplay the existence of various nationalisms and controversial readings of history, and to counteract potential heritage dissonance between the ethno-political groups. More than anything else it should be seen as an example of how heritage is not only a glance at the past but a projection into the future. In the evaluation document, the ICOMOS description of the 1992–1995 war was minimal, and did not even identify the different parties to the war. Neither did the description of the post-war situation of the city reveal any of the political tensions existing between the ethnic communities, or the actual division of the urban community along ethnic lines. In addition, neither the present ethnic composition of the city nor the historical formation of these communities was explained.[161] After its first visits to Mostar in 1999 and 2000, ICOMOS nevertheless felt the need to draw the attention of the State Party to the fact that the nomination should have the approval of both the Muslim and Croat communities. At a meeting organized at the time of the ICOMOS on-site visit, no representatives of the Croat community were present. Despite this fact, in 2000 ICOMOS recommended the inclusion of Mostar in the World Heritage List with a recommendation that the two communities should cooperate in the protection of the Old Town.[162] When in the same year the nomination was discussed and ultimately deferred by the World Heritage Committee, the reason for this deferral concerned the threat to the site due to uncontrolled building. The absence of support for the nomination by the Croat community was no longer addressed by the Committee or by ICOMOS in its final evaluation in 2005.[163] Apart from being a case in which multiculturalism was forced to have a single meaning, Mostar also shows how, even during the recent implementation of the Convention, nominating a World Heritage site has been possible without the full support of the local community. Finally, it should be noted that the nomination dossier for Bosnia Herzegovina, under the title "Mostar 1990–1996," was actually more voluble concerning the wartime and post-war political situation than the evaluation by ICOMOS.[164] In States Parties' nomination documents "old injuries and atrocities are often smoothed over."[165] The example of Mostar shows that ICOMOS, too, has been willing to smooth over old injuries.

The above examples suggest that a certain discomfort in producing narratives of negative heritage can be perceived in the ICOMOS evaluations as well.

This discomfort does not only limit itself to cities associated with recent traumatic wars like Warsaw and Mostar. All in all, the ICOMOS evaluations make only occasional reference to past urban ills. Most importantly, they fail to discuss cities in light of multiple and also conflicting interpretations. To provide one further example, we may look at how in its evaluation ICOMOS dealt with Zanzibar's difficult past with slavery. Through the use of criterion vi Zanzibar was described as being of "great symbolic importance" in the suppression of slavery. Elsewhere in the document, however, slavery was accorded only a few passing and neutral references: the trading of slaves by the Portuguese; the slave trade assuming large proportions in the later eighteenth century and its dislocation in the early nineteenth century; and the Anglican cathedral being built on the site of the last slave market. Notably, and in close congruence with the overall concentration on great-power histories, none of the history of slavery from the point of view of its human aspects was discussed as part of the description of the site. Instead, the focus of the description was placed on architectural history and the emergence of the so-called Swahili house.[166] This kind of fact-listing architectural description serves to promote neutral narratives of the past.

Naming and taming World Heritage cities

I will end this chapter by briefly discussing the labeling of World Heritage cities. Toponyms provide one important means of interpreting history: through naming streets, urban areas and even entire cities, societies utilize the past as a cultural, political and social resource.[167] In the World Heritage framework, however, the naming of cities has been conspicuously neutral and apolitical – this is visible in that most urban sites are called "historic centers," "historic cities" or "old towns." Very few urban World Heritage names actually openly advocate interpretations of past and present. A few labels highlight a specific limited historical period of significance (conversely concealing others periods). These include, for example, "Ferrara, City of the Renaissance and its Po Delta," the "Hanseatic Town of Visby," the "Medieval Town of Torun," and the "Late Baroque Towns of the Val di Noto." A conscious reversal of a representational regime in this sense was the change of the name of "Islamic Cairo," inscribed on the World Heritage List in 1979, into "Historic Cairo" in 2007 at the request of the Egyptian authorities. The outlined objective was to represent Cairo in a wider temporal perspective, "as the site encompasses monuments from other different eras."[168] Another incentive perhaps was a wish to move from a religiously focused label towards more neutral naming. The title "Stone Town of Zanzibar" is the only one which refers to the construction material traditionally used in the city. A few World Heritage titles highlight the monumental nature of the city ("Renaissance Monumental Ensembles of Úbeda and Baeza")[169] or underline urban stability and even museum-like characteristics ("Museum-City of Gjirokastra"), thus reflecting a particular attitude toward urban conservation. Furthermore, during the past decade it has become more common to refer to a particular type of city, for example "Provins, Town of Medieval Fairs" or "Liverpool – Maritime Mercantile City." These names are not only more informative and more concerned with thematic representation of heritage but also aim to make a projection into the future: by summing up the core of outstanding universal value for the cities in question they indicate what should be protected in their contexts.

There is only one former colonial city on the List, the "Colonial City of Santo Domingo," the title of which makes a direct reference to the colonial era. The twin-labeling of the "Portuguese City of Mazagan (El Jadida)" and "Medina of Essaouira (formerly Mogador)" also reveals a potentially controversial colonial heritage.[170] All other former colonial cities inscribed on the World Heritage List before 2008, are neutrally named as historic centers. This type of naming avoids highlighting controversial and one-sided interpretations of the past; it thus meshes with the overall aspiration of UNESCO for consensual pasts. Moreover, it shows perhaps the stakeholders' reluctance to take an active part in postcolonial debates. Nonetheless, it should be noted that, when looking at the wider descriptions of most of these cities as part of ICOMOS evaluations, even though apolitically named, the label and typology 'colonial city' is widely used to prescribe a meaning, and colonial history and architecture are frequently almost the sole focus of the articulation of outstanding universal value. Instead of multiple pasts, these discourses point towards a singular interpretation. One of the many examples is Lima, the outstanding universal value of which was justified in reference to "a Spanish colonial town of great political, economic and cultural importance in Latin America" and without any reference to its post-colonial history and heritage.[171] The more recent urban inscriptions from Latin America, most importantly those concerning Valparaíso and Cienfuegos, suggest, however, an altering perspective towards the inclusion of post-independence histories and heritages.[172]

It is not just concerning colonial cities that ICOMOS has been keen to label cities as part of its wider descriptions of outstanding universal value. Giving a name and a category is quite natural; it should be recalled, however, as was suggested earlier when chronological representation was discussed, that this form of labeling may also contribute to producing certain kinds of cities in terms of their future restoration, planning and representation. After all, there is a multitude of labels that could be used. As Anthony King has pointed out, at least 20 to 25 commonly used and a dozen less commonly used frames of reference for classifying cities or the quality of urbanism, "each suggesting an overriding variable," may be distinguished in the academic literature in English. These variables include, among others, technology (preindustrial, industrial, postindustrial), mode of production (capitalist, socialist, colonial, postcolonial), urban function (capital, administrative, commercial, industrial), national and cultural identity, ethnicity, religion, the level of economic development (rich, Third World), geographic scale (local, regional, national, international, global), size, chronology and location (continent or geographic location). What follows from this is that cities can be labeled in almost countless ways, and individual cities may fall into multiple categories in relation to one variable. Moreover, as King points out, although most systems of classification are relevant within the context of the majority of cities, there are some that have a more restricted use. For instance, religion is seen as an applicable attribute concerning 'Islamic cities' – this applies also to World Heritage – whereas in the case of 'Christian cities' it usually is not.[173] When a city is labeled, for instance, as a "Nordic city constructed in wood," as with the World Heritage site Old Rauma, only two variables are brought into focus – location/geographic scale and construction material. Many other aspects could be discussed as well. Proposing a category always involves selection.

Notes

1 Harrison, *Heritage: Critical Approaches*, 107–110; John E. Tunbridge, "Whose heritage to conserve? Cross-cultural reflections on political dominance and urban heritage conservation," in *The Heritage Reader*, ed. Graham Fairclough et al., (London: Routledge, 2008), 235–244; Tony Bennett, *The Birth of the Museum: History, Theory, Politics* (London: Routledge, 1995).

2 Hall, "Whose Heritage," 223; Dicks, *Culture*, 144–154; Joanne Maddern, "Huddled masses yearning to buy postcards: the politics of producing heritage at the Statue of Liberty–Ellis Island National Monument," *Current Issues in Tourism* 7: 4 (2004): 303–314.

3 UNESCO, World Heritage Committee, Eleventh Session (Paris, 7–11 December 1987), Report, SC-87/CONF.005/9, Paris, 20 January 1988) p. 11, http://whc.unesco.org/archive/1987/sc-87-conf005-9_e.pdf. United States, Canada and India opposed the inscription of Brasilia on the basis of paragraph 29 of the Operational Guidelines.

4 UNESCO, Bureau of the World Heritage Committee, Fifth Session (Paris, 4–7 May 1981), Report of Rapporteur, CC-81/CONF.002/4, Paris, 20 July 1981, p. 8, http://whc.unesco.org/archive/1981/cc-81-conf002-4e.pdf.

5 See, for example, UNESCO, Bureau of the World Heritage Committee, Tenth Session (Paris, 16–19 June, 1986), Report of Rapporteur, CC-86/CONF.001/11, Paris, 15 September 1986, http://whc.unesco.org/archive/1986/cc-86-conf001-11e.pdf; Pressouyre, *The World Heritage Convention*, 26.

6 UNESCO, World Heritage Committee, Eighth Session (Buenos Aires, Argentina, 29 October–2 November 1984), Report of the Rapporteur, SC/84/CONF.004/9, Buenos Aires, 2 November 1984, p. 5, http://whc.unesco.org/archive/1984/sc-84-conf004-9e.pdf; *Operational Guidelines*, January 1987, paragraph 29.

7 Pressouyre, *The World Heritage Convention*, 26–27.

8 Jokilehto et al., *The World Heritage List: Filling the Gaps*, 48–71.

9 Modern heritage properties on the World Heritage List (as at July 2006), accessed April 8, 2015, http://whc.unesco.org/uploads/activities/documents/activity-38-2.pdf.

10 ICOMOS evaluation for the nomination of the World Heritage property, "Housing Estates in Berlin," Germany, No 1230, 11 March 2008.

11 *Operational Guidelines*, 8 July 2015, Annex 3, paragraph 15.

12 ICOMOS evaluation for the nomination of the World Heritage property, "La Plata," Argentina, No 979, 21 January 2007. UNESCO, World Heritage Committee, Thirty-First Ordinary Session, Christchurch, New Zealand, 23 June – 2 July, Evaluations of cultural properties, WHC-07/31.COM/INF.8B1, p. 251–258, http://whc.unesco.org/archive/2007/whc07-31com-inf8b1e.pdf.

13 ICOMOS evaluation for the nomination of the World Heritage property, "Siena," Italy, No 717, September 1995.

14 Munasinghe, *Urban Conservation and City Life*, 49–50; G. J. Ashworth, "The conserved European city as cultural symbol: the meaning of the text," in *Modern Europe: Place, Culture and Identity*, ed. Brian Graham, 261–286 (London: Arnold, 1998), 269.

15 Koponen, Olli-Paavo, *Täydennysrakentaminen. Arkkitehtuuri, historia ja paikan erityisyys*. Tampere: Tampere University of Technology, 2006.

16 Jokilehto et al., *The World Heritage List: Filling the Gaps*, 54.
17 ICOMOS evaluation for the nomination of the World Heritage property, "Mostar," Bosnia and Herzegovina, No 946 rev, April 2005.
18 ICOMOS evaluation for the nomination of the World Heritage property, "Valletta," Malta, No 131, 1980.05. In another part of the document, though, the positive influence of nineteenth- and twentieth-century English architecture is mentioned.
19 ICOMOS evaluation for the nomination of the World Heritage property, "Strasbourg, Grand Ile," France, No 495, May 1988. For the more recent period, see, ICOMOS evaluation for the nomination of the World Heritage property, "Mantua and Sabbioneta," Italy, No 1287, 11 March 2008.
20 See, for example, ICOMOS evaluation for the nomination of the World Heritage property, "Cairo," Egypt, No 89, 1979.
21 ICOMOS evaluation for the nomination of the World Heritage property, "Florence," Italy, No 174, May 1982.
22 ICOMOS evaluation for the nomination of the World Heritage property, "Verona," Italy, No 797rev, September 2000.
23 The then-inscribed eight urban sites (some of them twin- or even multiple-city inscriptions) included the Historic Centers of Stralsund and Wismar, the Historic Inner City of Paramaribo, the Late Baroque Towns of the Val di Noto (South-Eastern Sicily), the Citadel, the Ancient City and Fortress Buildings of Derbent, the Historic Quarter of the Seaport City of Valparaíso, the Jewish Quarter and St Procopius' Basilica in Třebíč, the Renaissance Monumental Ensembles of Úbeda and Baeza, and the White City of Tel Aviv – the Modern Movement.
24 ICOMOS evaluation for the nomination of the World Heritage property, "Valparaíso," Chile, No 959rev, March 2003.
25 ICOMOS evaluation for the nomination of the World Heritage property, "Stralsund and Wismar," Germany, No 1067, January 2002.
26 ICOMOS evaluation for the nomination of the World Heritage property, "Paramaribo," Suriname, No 940rev, April 2002.
27 ICOMOS evaluation for the nomination of the World Heritage property, "Tel Aviv," Israel, No 1096, March 2003.
28 ICOMOS evaluation for the nomination of the World Heritage property, "Corfu," Greece, No 978, 11 March 2007. For other rare cases see ICOMOS evaluation for the nomination of the World Heritage property, "Salzburg," Austria, No 784, October 1996; ICOMOS evaluation for the nomination of the World Heritage property, "Diamantina," Brazil, No 890, September 1999; ICOMOS evaluation for the nomination of the World Heritage property, "Edinburgh," United Kingdom, No 728, September 1995.
29 ICOMOS evaluation for the nomination of the World Heritage property, "Karlskrona," Sweden, No 871, October 1998; ICOMOS evaluation for the nomination of the World Heritage property, "Tallinn," Estonia, No 822, September 1997.
30 ICOMOS evaluation for the nomination of the World Heritage property, "Bordeaux," France, No 1256, 21 January 2007.
31 ICOMOS evaluation for the nomination of the World Heritage property, "Liverpool," United Kingdom, No 1150, March 2004.
32 Ioannis Poulios, "Moving beyond a values-based approach to heritage conservation," *Conservation and Management of Archaeological Sites* 12: 2 (2010): 174.
33 The other European capital, Brussels, is represented in the World Heritage List by La Grand-Place. Because of this monumental delimitation of the site, it is not considered a World Heritage city in the framework of this research.
34 Pressouyre, *The World Heritage Convention*, 35.
35 ICOMOS evaluation for the nomination of the World Heritage property, "San Miguel and the Sanctuary of Atotonilco," Mexico, No 1274, 11 March 2008.
36 Jeffry M. Diefendorf and Janet Ward, "Introduction: Transnationalism and the German City," in *Transnationalism and the German City*, ed. Jeffry M. Diefendorf

and Janet Ward (New York: Palgrave Macmillan, 2014), 1. See also Deborah Cohen and Maura O'Connor, "Introduction: comparative history, cross-national history, transnational history – definitions," in *Comparison and History. Europe in Cross-National Perspective*, ed. Deborah Cohen and Maura O'Connor (New York: Routledge, 2004), xiii. For a transnational approach in urban history, see Nicholas Kenny and Rebecca Madgin, "'Every Time I Describe a City': Urban History as Comparative and Transnational Perspective," in *Cities Beyond Borders: Comparative and Transnational Approaches to Urban History*, ed. Nicholas Kenny and Rebecca Madgin (Farnham: Ashgate, 2015), 3–23.

37 Omland, "The ethics," 250.
38 ICOMOS evaluation for the nomination of the World Heritage property, "Úbeda et Baeza," Espagne, No 522, September 1989. World Heritage Centre, World Heritage Sites, Nomination files (1978–1999), Rejected (CD in author's possession).
39 ICOMOS evaluation for the nomination of the World Heritage property, "Úbeda-Baeza," Spain, No 522rev bis, March 2003.
40 ICOMOS evaluation for the nomination of the World Heritage property, "Potosi," Bolivia, No 420, April 1987.
41 Labadi, *UNESCO*, 72.
42 ICOMOS evaluation for the nomination of the World Heritage property, "Stralsund and Wismar," Germany, No 1067, January 2002.
43 Graham, Ashworth and Tunbridge, *A Geography of Heritage*; Omland, "The ethics," 250. For global history, see Mazlish, Bruce, "An introduction to global history," in *Conceptualizing Global History*, edited by Bruze Mazlish and Ralph Buultjens, 1–24. Boulder: Westview Press, 1993.
44 ICOMOS evaluation for the nomination of the World Heritage property, "Macao," China, No 1110, April 2005.
45 ICOMOS evaluation for the nomination of the World Heritage property, "Guanajuato," Mexico, No 482, September 1988; ICOMOS evaluation for the nomination of the World Heritage property, "Potosi," Bolivia, No 420, April 1987.
46 ICOMOS evaluation for the nomination of the World Heritage property, "Liverpool," United Kingdom, No 1150, March 2004.
47 ICOMOS evaluation for the nomination of the World Heritage property, "The Canal Area of Amsterdam," Netherlands, No 1349, 17 March 2010.
48 See, for example, ICOMOS evaluation for the nomination of the World Heritage property, "Valletta," Malta, No 131, 1980.05; ICOMOS evaluation for the nomination of the World Heritage property, "Jerusalem," nominated by Jordan, No 148, 1981; ICOMOS evaluation for the nomination of the World Heritage property, "Santiago de Compostcla," Spain, No 347, 1985; ICOMOS evaluation for the nomination of the World Heritage property, "Sana'a," Yemen, No 385, April 1986; ICOMOS evaluation for the nomination of the World Heritage property, "Kairouan," Tunisia, No 499, May 1988.
49 See, for example, ICOMOS evaluation for the nomination of the World Heritage property, "Angra do Heroismo," Portugal, No 206, June 1983; ICOMOS evaluation for the nomination of the World Heritage property, "Cartagena," Colombia, No 285, May 1984; ICOMOS evaluation for the nomination of the World Heritage property, "Island of Mozambique," Mozambique, No 599, May 1991; ICOMOS evaluation for the nomination of the World Heritage property, "Valparaíso," Chile, No 959 rev, March 2003; ICOMOS evaluation for the nomination of the World Heritage property, "Macao," China, No 1110, April 2005.
50 ICOMOS evaluation for the nomination of the World Heritage property, "Leningrad," Soviet Union, No 540, April 1990.
51 ICOMOS evaluation for the nomination of the World Heritage property, "Zanzibar," Tanzania, No 173rev, September 2000.
52 ICOMOS evaluation for the nomination of the World Heritage property, "Panamá," Panama, No 790, September 1997.

53 ICOMOS evaluation for the nomination of the World Heritage property, "Mostar," Bosnia and Herzegovina, No 946 rev, April 2005; ICOMOS evaluation for the nomination of the World Heritage property, "Warsaw," Poland, No 30, June 6, 1978.

54 ICOMOS evaluation for the nomination of the World Heritage property, "Prague," Czechoslovakia, No 616, October 1992; ICOMOS evaluation for the nomination of the World Heritage property, "Salzburg," Austria, No 784, October 1996; ICOMOS evaluation for the nomination of the World Heritage property, "Vienna," Austria, No 1033, September 2001.

55 ICOMOS evaluation for the nomination of the World Heritage property, "Alcalá de Henares," Spain, No 876, October 1998; ICOMOS evaluation for the nomination of the World Heritage property, "Prague," Czechoslovakia, No 616, October 1992.

56 ICOMOS evaluation for the nomination of the World Heritage property, "Brugge," Belgium No 996, September 2000.

57 ICOMOS evaluation for the nomination of the World Heritage property, "Leningrad," Soviet Union, No 540, April 1990.

58 ICOMOS evaluation for the nomination of the World Heritage property, "Venice and its Lagoon," Italy, No 394, May 1987.

59 MacCannell, *The Tourist.*

60 Maddern, "Huddled masses," 312.

61 Labadi, "Representations," 160.

62 ICOMOS evaluation for the nomination of the World Heritage property, "Guimarães," Portugal, No 1031, September 2001.

63 ICOMOS evaluation for the nomination of the World Heritage property, "Arequipa," Peru, No 1016, September 2000.

64 ICOMOS evaluation for the nomination of the World Heritage property, "Morelia," Mexico, No 585, November 1991.

65 ICOMOS evaluation for the nomination of the World Heritage property, "Sintra," Portugal, No 723, September 1995.

66 See, for example, Koshar, *Germany's Transient Pasts.*

67 Wendy Beck, "Narratives of World Heritage in Travel Guidebooks," *International Journal of Heritage Studies* 12: 6 (2006): 529–530.

68 See, for example, ICOMOS evaluation for the nomination of the World Heritage property, "San Marino," San Marino, No 1245, 11 March 2008.

69 Pressouyre, *The World Heritage Convention*, 35–36.

70 ICOMOS evaluation for the nomination of the World Heritage property, "Trakai," Lithuania, No 1176, April 2005. UNESCO, World Heritage Committee, Twenty-Ninth Session, Durban, South Africa, 10–17 July 2005, Evaluations of cultural properties, WHC.05/29.COM/INF.8B.1 (2005), p. 135–141, http://whc.unesco.org/archive/2005/whc05-29com-inf08B1e.pdf.

71 ICOMOS evaluation for the nomination of the World Heritage property, "Old City and Ramparts of Alanya," Turkey, No 1354, 10 March 2011.

72 Ibid., 141. See also, for example, ICOMOS evaluation for the nomination of the World Heritage property, "Tripoli," Lebanon, No 298, December 2, 1983. World Heritage Centre, World Heritage Sites, Nomination files (1978–1999), Rejected (CD in author's possession); ICOMOS evaluation for the nomination of the World Heritage property, "Gerona," Spain, No 519, September 1989. World Heritage Centre, World Heritage Sites, Nomination files (1978–1999), Rejected (CD in author's possession); ICOMOS evaluation for the nomination of the World Heritage property, "Taal," Philippines, No 501, September 1989. World Heritage Centre, World Heritage Sites, Nomination files (1978–1999), Rejected (CD in author's possession).

73 Titchen, *On the Construction*, 90–94.

74 J. Carman, *Against Cultural Property: Archaeology, Heritage and Ownership* (London: Duckworth, 2005).

75 Richard Rodger and Roey Sweet, "The changing nature of urban history." *History in Focus*, February 2008, accessed March 3, 2015, www.history.ac.uk/ihr/Focus/City/articles/sweet.html#t14.

76 H. J. Dyos, "Editorial." *Urban History Yearbook* (1974), 5.

77 For a global overview, see, Peter Clark, ed., *Cities in World History* (Oxford: Oxford University Press, 2013).

78 ICOMOS evaluation for the nomination of the World Heritage property, "Derbent," Russian Federation, No 1070, March 2003.

79 For the origins of the urban history discipline in social and economic history, and cultural turn since the 1980s, see Simon Gunn, "The spatial turn: changing histories of space and place," in *Identities in Space: Contested Terrains in the Western City since 1850*, ed. Simon Gunn and Robert J. Morris (Aldershot: Ashgate, 2001), 1–14.

80 ICOMOS evaluation for the nomination of the World Heritage property, "Brugge," Belgium, No 996, September 2000.

81 ICOMOS evaluation for the nomination of the World Heritage property, "Provins," France, No 873, October 1998. World Heritage Centre, World Heritage Sites, Nomination files (1978–1999), Rejected (CD in author's possession).

82 ICOMOS evaluation for the nomination of the World Heritage property, "Provins," France, No 873rev, September 2001.

83 ICOMOS evaluation for the nomination of the World Heritage property, "São Luís," Brazil, No 821, September 1997.

84 For discussion concerning urban light and darkness, see Chris Otter, *The Victorian eye: a political history of light and vision in Britain, 1800–1910* (Chicago: University of Chicago Press, 2008).

85 ICOMOS evaluation for the nomination of the World Heritage property, "The Canal Area of Amsterdam," Netherlands, No 1349, 17 March 2010.

86 For a similar account concerning national nomination dossiers see Labadi, "Representations."

87 ICOMOS evaluation for the nomination of the World Heritage property, "Naples," Italy, No 726, September 1995.

88 ICOMOS evaluation for the nomination of the World Heritage property, "La Chaux-de-Fonds/Le Locle," Switzerland, No 1302, 10 March 2009.

89 ICOMOS evaluation for the nomination of the World Heritage property, "Liverpool," United Kingdom, No 1150, March 2004.

90 ICOMOS evaluation for the nomination of the World Heritage property, "Valparaíso," Chile No 959rev, March 2003.

91 See, however, ICOMOS evaluation for the nomination of the World Heritage property, "Provins," France, No 873rev, September 2001.

92 The World Heritage mining towns include Røros (Norway, 1980), Ouro Preto (Brazil, 1980), Potosi (Bolivia, 1987), Guanajuato (Mexico, 1988), Goslar (Germany, 1992), Zacatecas (Mexico, 1993), Banska Stiavnica (Slovakia, 1993), Kutná Hora (Czech Republic, 1995) and Hallstatt (Austria, 1997).

93 For industrial nominations and tentative nominations excluding the urban structure, see ICOMOS evaluation for the nomination of the World Heritage property, "Zollverein," Germany, No 975, September 2001; and "Tentative Lists, United Kingdom of Great Britain and Northern Ireland, Manchester and Salford," accessed June 6, 2011, http://whc.unesco.org/en/tentativelists/1316/.

94 ICOMOS evaluation for the nomination of the World Heritage property, "Baku," Azerbaijan, No 958, September 2000. See also ICOMOS evaluation for the nomination of the World Heritage property, "Verona," Italy, No 797rev, September 2000.

95 ICOMOS evaluation for the nomination of the World Heritage property, "Riga," Latvia, No 852, September 1997.

96 ICOMOS evaluation for the nomination of the World Heritage property, "Dresden Elbe Valley," Germany, No 1156, March 2004.

97 Peter Clark, *European Cities and Towns 400–2000* (New York: Oxford University Press, 2009), 248–250.

98 Ibid.

99 John K. Walton and Jason Wood, "Reputation and regeneration: history and the heritage of the recent past in the re-making of Blackpool," in *Valuing Historic Environments*, ed. Lisanne Gibson and John Pendlebury (Surrey: Ashgate, 2009), 115–137.

100 Tentative Lists, United Kingdom of Great Britain and Northern Ireland, November 3, 2015, http://whc.unesco.org/en/tentativelists/state=gb.

101 ICOMOS evaluation for the nomination of the World Heritage property, "Innsbruck," Austria, No 1169, April 2005. UNESCO, World Heritage Committee, Twenty-ninth session, Durban, South Africa, 10–17 July 2005, Evaluations of cultural properties, WHC.05/29.COM/INF.8B.1 (2005), p. 87–91, http://whc.unesco.org/archive/2005/whc05-29com-inf08B1e.pdf.

102 Tentative Lists, Austria, Great Spas of Europe, November 3, 2015, http://whc.unesco .org/en/tentativelists/5930/.

103 ICOMOS evaluation for the nomination of the World Heritage property, "Assisi," Italy, No 990, September 2000.

104 ICOMOS evaluation for the nomination of the World Heritage property, "St George," United Kingdom, No 983, September 2000.

105 John Urry, *The Tourist Gaze*. 2nd edn. (Sage: London, 2002); Kirshenblatt-Gimblet, *Destination*; Dicks, *Culture*.

106 ICOMOS, *Cultural Tourism*. International Scientific Symposium (ICOMOS, 10th General Assembly: Sri Lanka, 1993).

107 ICOMOS evaluation for the nomination of the World Heritage property, "Třebíč," Czech Republic, No 1078, March 2003.

108 ICOMOS evaluation for the nomination of the World Heritage property, "Dresden Elbe Valley," Germany, No 1156, March 2004.

109 UNESCO, World Heritage Committee, Thirty-third session (Seville, Spain 22–30 June 2009), Report of the decisions, WHC-09/33.COM/20, Seville, 20 July 2009, p. 43–44, http://whc.unesco.org/archive/2009/whc09-33com-20e.pdf.

110 ICOMOS evaluation for the nomination of the World Heritage property, "Housing Estates in Berlin," Germany, No 1230, 11 March 2008.

111 ICOMOS evaluation for the nomination of the World Heritage property, "Ouro Preto," Brazil, No 124, May 1980.

112 ICOMOS evaluation for the nomination of the World Heritage property, "Morelia," Mexico, No 585, November 1991.

113 ICOMOS evaluation for the nomination of the World Heritage property, "Arequipa," Peru, No 1016, September 2000.

114 ICOMOS evaluation for the nomination of the World Heritage property, "Yaroslavl," Russian Federation, No 1170, April 2005; ICOMOS evaluation for the nomination of the World Heritage property, "Úbeda–Baeza," Spain, No 522rev bis, March 2003; ICOMOS evaluation for the nomination of the World Heritage property, "Noto," Italy, No 1024rev, January 2002.

115 ICOMOS evaluation for the nomination of the World Heritage property, "Noto," Italy, No 1024rev, January 2002.

116 Labadi, "Representations," 162–163; Labadi, *UNESCO*, 78–88.

117 Laurajane Smith, "Heritage, Gender and Identity," in *Ashgate Research Companion to Heritage and Identity*, ed. Brian Graham and Perter Howard (Abingdon: Ashgate, 2012), 159–162. For quotation see page 159.

118 ICOMOS evaluation for the nomination of the World Heritage property, "Zacatecas," Mexico, No 676, October 1993; ICOMOS evaluation for the nomination of the World Heritage property, "Guanajuato," Mexico, No 482, September 1988; ICOMOS

evaluation for the nomination of the World Heritage property, "Potosi," Bolivia, No 420, April 1987. See also ICOMOS evaluation for the nomination of the World Heritage property, "Visby," Sweden, No 731, September 1995; ICOMOS evaluation for the nomination of the World Heritage property, "Mauritanian towns," Mauritania, No 750, October 1996.

119 Hayden, Dolores, "The Power of Place Project: Claiming Women's History in the Urban Landscape," in *Restoring Women's History through Historic Preservation*, ed. Gail Lee Dubrov and Jennifer B. Goodman (Baltimore: The Johns Hopkins University Press, 2003), 199–213.

120 ICOMOS evaluation for the nomination of the World Heritage property, "Leningrad," Soviet Union, No 540, April 1990.

121 ICOMOS evaluation for the nomination of the World Heritage property, "Goiás," Brazil, No 993, September 2001.

122 ICOMOS evaluation for the nomination of the World Heritage property, "Tel Aviv," Israel, No 1096, March 2003.

123 ICOMOS evaluation for the nomination of the World Heritage property, "Verona," Italy, No 797rev, September 2000.

124 ICOMOS evaluation for the nomination of the World Heritage property, "Warsaw," Poland, No 30, June 6, 1978; ICOMOS evaluation for the nomination of the World Heritage property, "Brugge," Belgium, No 996, September 2000; ICOMOS evaluation for the nomination of the World Heritage property, "Mostar," Bosnia and Herzegovina, No 946 rev, April 2005.

125 ICOMOS evaluation for the nomination of the World Heritage property, "Lamu Old Town," Kenya, No 1055, September 2001. See also ICOMOS evaluation for the nomination of the World Heritage property, "Zansibar," Tanzania, No 173rev, September 2000.

126 Labadi, *UNESCO*, 85.

127 Cara Aitchison, "Heritage and nationalism: gender and the performance of power," in *Leisure/Tourism Geographies: Practices and Geographical Knowledge*, ed. David Crouch (London: Routledge, 1999), 65.

128 UNESCO, World Heritage Committee, Third session (Cairo and Luxor, 22–26 October 1979), Report of Rapporteur, CC-79/CONF.003/13, Paris, 30 November 1979, http://whc.unesco.org/archive/1979/cc-79-conf003-13e.pdf.

129 Labadi, *UNESCO*, 79. Concerning the regulation of the treatment of difference in cross-cultural manner, see Ulrich Beck, *The Cosmopolitan Vision* (Cambridge: Polity Press, 2006), 49.

130 ICOMOS evaluation for the nomination of the World Heritage property, "Ferrara," Italy, No 733, September 1995.

131 Labadi, *UNESCO*, 79–80.

132 Smith, "Heritage, Gender and Identity," 161.

133 Pierre Galland et al., *World Heritage in Europe Today* (Paris: UNESCO, 2016), 25.

134 ICOMOS evaluation for the nomination of the World Heritage property, "Leningrad," Soviet Union, No 540, April 1990.

135 ICOMOS evaluation for the nomination of the World Heritage property, "Potosi," Bolivia, No 420, April 1987.

136 ICOMOS evaluation for the nomination of the World Heritage property, "Lyons," France, No 872, October 1998. Concerning nomination dossiers, see Labadi, *UNESCO*, 87–88.

137 Barry Doyle, "A Decade of Urban History: Ashgate's Historical Urban Studies Series," *Urban History* 36: 3 (2009): 510.

138 See also Omland, "The ethics," 249; Labadi, "Representations," 161.

139 Sharon Macdonald, *Difficult Heritage: Negotiating the Nazi Past in Nuremberg and Beyond* (London: Routledge, 2009).

140 Meskell, "Negative heritage," 558.

141 Trinidad Rico, "Negative heritage: the place of conflict in World Heritage." *Conservation and Management of Archeological Sites* 10: 4 (2008): 347.
142 Omland, "The ethics," 252.
143 The politicization in association with difficult heritage was especially felt at the 1996 World Heritage Committee meeting on the occasion of the inscription of Genbaku Dome in Hiroshima. UNESCO, World Heritage Committee (Merida, Yucatan, Mexico, 2–7 December 1996), Twentieth session, Report, WHC-96/CONF.201/21, 10 March 1997, Annex V., http://whc.unesco.org/archive/1996/whc-96-conf201-21e.pdf.
144 Rico, "Negative heritage," 348. On the failure of the World Heritage List to represent heritage sites with negative meanings see also Meskell, "Negative heritage;" Jan Turtinen, "Globalising heritage – on UNESCO and the transnational construction of a World Heritage," *SCORE Rapportserie* No. 12, 2000, accessed February 2, 2012, www.score.su.se/polopoly_fs/1(2000).26651.1320939806!/200012.pdf.
145 Rico, "Negative heritage," 344.
146 ICOMOS evaluation for the nomination of the World Heritage property, "Warsaw," Poland, No 30, June 6, 1978.
147 Ibid.
148 ICOMOS evaluation for the nomination of the World Heritage property, "Mostar," Bosnia and Herzegovina, No 946 rev, April 2005.
149 ICOMOS evaluation for the nomination of the World Heritage property, "Sarajevo," Yugoslavia, No 388, April 1986. World Heritage Centre, World Heritage Sites, Nomination files (1978–1999), Rejected (CD in author's possession).
150 ICOMOS evaluation for the nomination of the World Heritage property, "Sarajevo," Bosnia and Herzegovina, No 851rev, 1999. World Heritage Centre, World Heritage Sites, Nomination files (1978–1999), Rejected (CD in author's possession).
151 ICOMOS evaluation for the nomination of the World Heritage property, "Baku," Azerbaijan, No 958, September 2000.
152 ICOMOS evaluation for the nomination of the World Heritage property, "Segovia," Spain, No 311 rev., November 1985; ICOMOS evaluation for the nomination of the World Heritage property, "Cordoba," Spain No 313bis, October 1994.
153 ICOMOS evaluation for the nomination of the World Heritage property, "Bardejov," Slovakia, No 973, September 2000.There is no Jewish population living in the city today.
154 Luda Klusáková, "Between urban and rural culture: public use of history and cultural heritage in building collective identities (1990–2007)," in *Being a Historian – Opportunities and Responsibilities – Past and Present*, ed. Sven Mörsdorf. A CLIOHRES-ISHA Reader II, www.cliohworld.net/docs/isha2_p.pdf, 207–208. Jewish heritage also played an important part in the inscriptions of Ferrara and Essaouira.
155 ICOMOS evaluation for the nomination of the World Heritage property, "Old City and Ramparts of Alanya," Turkey, No 1354, 10 March 2011. UNESCO, World Heritage Committee, 35th Session (UNESCO, June 2011), ICOMOS Evaluations of Nominations of Cultural and Mixed Properties, p. 307–316, http://whc.unesco.org/archive/2011/whc11-35com-inf.8B1e.pdf; ICOMOS evaluation for the nomination of the World Heritage property, "Jajce," Bosnia and Herzegovina, No 1294, 10 March 2009. UNESCO, World Heritage Committee, 33rd Session (UNESCO, 22–30 June 2009, Seville), ICOMOS Evaluations of Cultural Properties, WHC-09/33.COM/ INF.8B1, p. 112–120, http://whc.unesco.org/archive/2009/whc09-33com-inf8B1e.pdf.
156 Harrison, Rodney, "The politics of the past. Conflict in the use of heritage in the modern world," in *The Heritage Reader*, ed. Graham Fairclough et al. (London: Routledge, 2008), 183.
157 Dicks, *Culture*, 151, in reference to A. E. Coombes, "Inventing the 'postcolonial': hybridity and constituency in contemporary curating," *New Formations* 18 (1991): 39–52.

158 Emily Gunzburger Makaš, *Representing Competing Identities: Building and Rebuilding in Postwar Mostar, Bosnia-Herzegovina.* Unpublished Ph.D. Dissertation, Cornell University, 2007, 420.
159 The International Relations and Security Network, ETH Zurich, "Hopeful rebirth for Bosnia's divided Mostar," February 3, 2004, accessed June 3, 2014, www.isn.ethz.ch/isn/Current-Affairs/Security-Watch/Detail/?ots591=4888CAA0-B3DB-1461-98B9-E20 E7B9C13D4&lng=en&id=107315.
160 ICOMOS evaluation for the nomination of the World Heritage property, "Mostar," Bosnia and Herzegovina, No 946 rev, April 2005; Emily Gunzburger Makaš, "Rebuilding Mostar: international and local visions of a contested city and its heritage," in *On Location: Heritage Cities and Sites*, ed. D. F. Ruggles (New York: Springer, 2012), 165; Gunzburger Makaš, *Representing Competing Identities*, 404. Gunzburger Makaš, "Rebuilding Mostar," 158, explains that most of the Croat population of Mostar "subtly obstructed or simply ignored the whole reconstruction process."
161 ICOMOS evaluation for the nomination of the World Heritage property, "Mostar," Bosnia and Herzegovina, No 946 rev, April 2005.
162 ICOMOS evaluation for the nomination of the World Heritage property, "Mostar," Bosnia and Herzegovina, No 946, November 2000.
163 UNESCO, World Heritage Committee (Cairns, Australia, 27 November – 2 December 2000), Twenty-fourth session, Report, WHC.00/CONF.204/21, Paris, 16 February 2001, p. 51, http://whc.unesco.org/archive/2000/whc-00-conf204-21e.pdf; ICOMOS, No 946 rev, April 2005 (Mostar, Bosnia and Herzegovina).
164 Bosnia and Herzegovina, Nomination, "The Old City of Mostar," January 2005, accessed January 15, 2013, http://whc.unesco.org/uploads/nominations/946rev.pdf.
165 Turtinen, "Globalising heritage."
166 ICOMOS evaluation for the nomination of the World Heritage property, "Zanzibar," Tanzania, No 173rev, September 2000.
167 For discussion about urban naming see Jani Vuolteenaho and Lawrence D. Berg, "Towards critical toponymies," in *Critical Toponymies: The Contested Politics of Place Naming*, ed. Lawrence D. Berg and Jani Vuolteenaho (Surrey: Ashgate, 2009), 1–18; Yvonne Whelan, "Mapping Meanings in the Cultural Landscape," in *Senses of Places: Senses of Time*, ed. G. J. Ashworth and Brian Graham (Aldershot: Ashgate, 2005), 61–71.
168 UNESCO, World Heritage Committee, Thirty-First Session (Christchurch, New Zealand, 23 June – 2 July 2007), Item 8B of the Provisional Agenda: Nominations to the World Heritage List, WHC-07/31.COM/8B, Paris, 11 May 2007, p. 1, http://whc.unesco.org/archive/2007/whc07-31com-8be.pdf.
169 Other examples of World Heritage cities labeled by highlighting the monumental nature of the city include the Historical Complex of Split with the Palace of Diocletian (Yugoslavia, 1979), the Historic Monuments Zone of Querétaro (Mexico, 1996), and the Historic Monuments Zone of Tlacotalpan (Mexico, 1998).
170 See also Oliver Creighton, "Contested Townscapes: the Walled City as World Heritage," *World Archaeology* 39: 3 (2007): 347.
171 ICOMOS evaluation for the nomination of the World Heritage property, "Lima," Peru, No 500bis, November 1991. For discussion concerning Lima as a World Heritage site see Seppänen, *Global Scale.*
172 ICOMOS evaluation for the nomination of the World Heritage property, "Valparaíso," Chile, No 959rev, March 2003; ICOMOS evaluation for the nomination of the World Heritage property, "Cienfuegos," Cuba, No 1202, April 2005.
173 Anthony King, "Terminologies and types: making sense of some types of dwellings and cities," in *Ordering Space: Types in Architecture and Design*, ed. Karen A. Frank and Lynda H. Schneekloth (New York: Van Nostrand Reinhold, 1994), 139.

5 World Heritage cities

What urban futures?

It would be difficult to find a city, which offers in as limited an area as many unique aesthetic creations, whose influence on the evolution of art and architecture, has been as great.[1]

Of particular significance are two issues: a) the social organization of the communities through the Afocha and the Kebele administration, b) the close urban-rural linkages, which are also significant in the Harari language [...] Harar Jugol is considered to be of outstanding universal value having also exceptionally well preserved its social and physical inheritance.[2]

Rome, the center of the Roman Republic and the Roman Empire, later became the capital of the Christian world. The fortified town of Harar, located in the eastern part of Ethiopia, is the fourth holiest city of Islam. Both Rome and Harar Jugol are exceptional and impressive places. However, the above extracts from the ICOMOS statements on determining outstanding universal value for these two cities represent almost completely opposite extremes. They immediately point towards two things: firstly, that there can be highly diversified ways in which to express the outstanding universal value of a city, and, secondly, because these two statements represent different periods during the implementation of the World Heritage Convention that there has been a significant change in the World Heritage valuation discourse. What falls between these two extremes will be the focus of this chapter. The primary emphasis will be placed on how and to what extent the values articulated in reference to the outstanding universal value of cities have developed over the years in relation to the three separate but intertwined processes: the evolving perceptions of cultural heritage in society; the broadening spectrum of heritage values in professional usage; and the principles of integrated conservation, advocating the parallel consideration of conservation and development in the context of cities (as articulated in the Vienna Memorandum). To this end several themes will be explored, including monumentality, unity, harmony, setting, social and local community values, authenticity, continuity and change. In addition to conceptual change, what may be discerned in the ICOMOS discourse on cities is continuity, often to the point of an inability to move beyond the well-established Western-based rhetoric about a material, visual and stagnant city.

Urban monuments

Despite the fact that during the 1960s and 1970s the attention in professional conservation shifted towards everyday environments and area conservation,[3] at the early stage of the implementation of the World Heritage Convention a monument-centered approach to the evaluation and selection of cultural heritage was dominant. This monumental view was subject to criticism in the Global Strategy articulated in 1994, which proposed instead a more anthropological and holistic understanding of cultural heritage; however, states in their nomination dossiers even after that date have frequently used monumentality to present the nominated sites as national icons.[4] The intertwined notions of monumentality, grand scale and nation building make up the core of authorized discourses concerning heritage.[5] Urban heritage at least has the potential of offering broader perspectives on cultural heritage; it is thus important to begin this chapter by exploring the role that monumentality – in the meaning of a monument to something grand, unique, individual, isolated and eternal – has played as part of the ICOMOS' construction of outstanding universal value.

The early ICOMOS evaluations imposed a rather mechanistic formula and language, be that in relation to cities or to more conventional monuments, such as European cathedrals, palaces or castles. Often, the language used in the valuation of these conventional monuments was applied with very little adjustment to the description of cities as well. Thus cities, in their entirety, could be conceived of as monuments and works of art, as "unique artistic and urbanistic master piece[s],"[6] or sometimes even in very straightforward monumental language as "World Heritage Monument[s]."[7] Furthermore, it was common to describe cities by describing their individual monuments. When simplified a little, it could be said that the higher the concentration of monuments, the higher the World Heritage value was regarded, as exemplified in the case of Valletta: "The total of 320 historic monuments which exists within a confined area of 55 hectares is among the most strongly concentrated of this nature in the world."[8] The same approach was also applied to modern Brasilia. In its evaluation ICOMOS emphasized the monumental nature of the city, its plan and its public buildings, and the great ingenuity of the two masters – Lucio Costa and Oscar Niemeyer – who had created it.[9] The uniqueness or representativeness of a city was essentially seen as something based on the uniqueness or representativeness of its key monuments.[10] Sometimes the contrast created by the existence of both monumental and simpler vernacular architectural features was highlighted as a basis for determining outstanding universal value, the main emphasis, however, being placed on monumentality:

> One of the twenty or so Baroque churches [...] appears unexpectedly as one turns a corner – the studied refinement of their décor contrasting with the charming simplicity of the houses, which are painted in vivid colours or faced with ceramic tiles.[11]

It is also possible to find this monumental approach when looking at urban nominations for which ICOMOS recommended rejection. On average, the rejected cities may be characterized as less monumental in the conventional use of the term (see Appendix 2), even though, at the same time, it ought to be noted that the lacking outstanding universal value of these cities was usually not explicitly defined on grounds of missing monumental value. Nevertheless, in the context of its negative statement concerning the outstanding universal value of Plovdiv ICOMOS noted that "Plovdiv is a city richer in history than it is in monuments."[12] The World Heritage Committee, at the time of the official rejection of Plovdiv's nomination, considered that it was "difficult at this stage to include urban sites on the list for their vernacular architecture and that the problems concerning the types of towns characteristic of the different regions of the World would first have to be clarified."[13] These considerations make visible the early stage coupling of monumentality with outstanding universal value in contrast to vernacularity equated with regional and local values. We may find change over the years in this respect: for instance, in its negative statement concerning Alanya in 2011, ICOMOS noted that "the value of Alanya as an old city cannot derive from the value of individual monuments" alone.[14]

The monumental approach applied to cities cannot be discussed without making reference to the pronouncedly Eurocentric framework of analysis applied with regard to the identification of World Heritage. The urban category within the World Heritage List may appear "truly global" in terms of geographical representation,[15] but, perhaps more than anything else, this has been because many other categories have been even more unevenly distributed in the context of the World Heritage List. In congruence with the criticized over-representation of European heritage on the World Heritage List, a World Heritage city has been first and foremost a European city, and to a second, lesser degree a Latin American city (Figure 5.1), in fact often represented by a European colonial city.[16] Especially during the early implementation of the World Heritage Convention, monumental character, typical of many of the most valued European historic cities, was a standard which ICOMOS generally assumed to also apply to cities from other parts of the world. If the monumental condition was not fulfilled, a justification in terms of this representational 'failure' was expected.

On the other hand, it should also be noted that the division of rejected urban nominations, when defined in relation to UNESCO regions, rather faithfully follows the geographical distribution of accepted nominations. About 60 percent of the World Heritage nominated cities that ICOMOS recommended for rejection are located in Europe and North America and close to another 20 percent in Latin America and the Caribbean. This suggests that the low number of World Heritage cities located in other regions, especially in Africa, cannot be attributed to their high rejection percentage but to both their lower nomination rate and to their 'disappearance' into the system without an official rejection (deferral or referral back to the State Party), as discussed in Chapter 1.

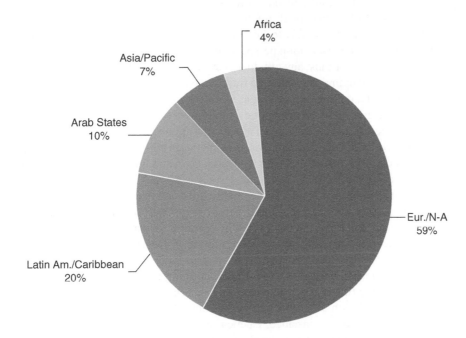

Figure 5.1 World Heritage cities (187) by UNESCO regions, 1978–2011

To an important degree, the interplay of the monumental and Eurocentric views was related to the use and scope of criteria i and v. Even though in 1995 the reference to "unique artistic achievement" was removed from the description of criterion i as a direct response to the adoption of the Global Strategy, the other monumental connotations of creativity, masterpiece and genius were nevertheless retained. Between the years 1978 and 2011 criterion i was applied in association with 41 cities, hence a little over a fifth of the total of 187 cities included in the research data (Appendix 3). With the exception of a few Arabic cities (Damascus, Cairo, Marrakesh and Kairouan), criterion i was almost solely used with regard to European cities. While applied 15 times out of the total of 41 in reference to Italian or Spanish cities, not a single association with the concept of a masterpiece was made in the context of African cities, and only one in relation to an Asian city, Samarkand. However, what is notable is that between 2002 and 2011 criterion i was accepted by ICOMOS only once in reference to urban nominations (The Canal Area of Amsterdam, 2011), even though it was still occasionally proposed by states.[17] This shows that ICOMOS and the World Heritage Committee have more recently wanted to move away from the most monumental and fine arts understanding of what makes up urban World Heritage.[18]

Even though the majority of early urban inscriptions fell into the monumental category, it should be pointed out that there were also notable exceptions to this rule, such as the inscriptions of Ghadamés (1986) or Mozambique (1991), and those concerning the Nordic wooden towns, Røros (1979), Bryggen (1980) and Rauma (1991). In fact (as noted in Chapter 2), criterion v had been introduced in order for the List to be able to host such cultural testimonies that, from the established Western conservation perspective, appeared vernacular. One might be inclined to think that almost any city that is awarded World Heritage status would constitute an outstanding example of a traditional human settlement and land-use representative of a culture or cultures, as stated in criterion v. The reverse is true, however, since criterion v has been one of the less-frequently used criteria in the context of cities. Between 1978 and 1993, it was applied in reference to 22 cities out of the 82 inscribed during that period, and between 1994 and 2011 in reference to 25 cities out of a total of 105 (see Appendix 2). Although criterion v has been used in the context of some Western cities, such as Rhodes, Visby or Dresden, its key ideas of a traditional human settlement and irreversible change have primarily been associated with non-Western urban heritage. What is striking is how this criterion was used before 1994 most often in reference to Islamic cities considered as "having become vulnerable through the impact of irreversible socio-economic changes."[19] It was more common for ICOMOS to think of Islamic cities as representative of traditional architecture, thus suggesting a framing of this heritage in juxtaposition to the colonial "modern" city.[20] In the latter part of the 1990s the application of criterion v became somewhat more differentiated and geographically more dispersed. For instance, Cuenca, by using criterion v, was described in a generalized manner as "an outstanding example of a planned inland Spanish colonial city."[21] In 2000–2011 the criterion was used only seven times, which suggests that the importance of this criterion in the context of World Heritage cities was also decreasing. This seems somewhat surprising in light of the Global Strategy's emphasis on non-monumentality.

It was precisely criterion v and its notion of traditional human settlements that allowed the inclusion in the World Heritage List of a small group of conventionally less monumental cities before the adoption of Global Strategy in 1994. The role of these cities in the framework of the World Heritage List should nevertheless be seen as the role of 'the other,' providing contrast to the monumental architecture considered to be the prime concern. Moreover, what should be pointed out is that often a similar monumental logic characterized the discursive representation of these less monumental cities than with those that might be considered conventionally monumental."[22]

When looking at the ICOMOS evaluations from the late 1990s up until 2011, it becomes apparent that the most straightforward monumental approach was abandoned. Increasing emphasis was placed on such aspects as the overall urban fabric and the spatial development of a city. Also some of the later inscriptions of earlier rejected nominations should be seen as a direct outcome of the Global Strategy's criticism towards a universalist monumental approach (Appendix 2). In the cases of Zanzibar, Vigan and Cidade Velha of Ribeira Grande, all nominations that were rejected prior to 1994, ICOMOS subsequently made a reassessment concerning their outstanding universal value and turned positive towards their inscription on the World Heritage List. The negative ICOMOS evaluation concerning Vigan in 1989 shows how ICOMOS at that time wanted to follow a strict policy of the "best of the best." The outstanding universal value was seen to be perceptible only in the global perspective of Spanish colonial towns.[23] In 1999, as ICOMOS was engaged in the second evaluation of Vigan after its re-nomination by the Philippine authorities, the organization acknowledged that its earlier comparison with Spanish colonial towns in the Caribbean had been an invalid one. Advocating the regionalization of the definition of outstanding universal value after the adoption of the Global Strategy, ICOMOS now noted that "historic towns should be evaluated in a regional context rather than globally."[24]

The 1998 *Management Guidelines for World Heritage Sites* made explicit the importance of non-monumental urban architecture in consideration of World Heritage inscribed cities: "(it) is these structures and urban spaces in which the life of the town has evolved that distinguish the concept of historic town from a group of monuments."[25] The more recent ICOMOS evaluations repeat this statement rather meticulously. Hence, "the merit of Arequipa architecture is not limited to the grandeur of its religious monuments. It is also in the profusion of dignified *casonas,* characteristic well proportioned vernacular houses."[26] Some of this respect for non-monumental "anonymous architecture" is often belied, however, by a simultaneous long and detailed list and description of the most important monumental buildings of a city, often placed hierarchically above the non-monumental aspects.[27] The importance of "a large number of ancient buildings of monumental importance" for the outstanding universal value of cities also continues to be highlighted in the Operational Guidelines.[28] This suggests that, even though significantly modified, monumentality continues to serve as a rhetorical guarantee of representing World Heritage value: the more evidently monumental, the more outstandingly universal.

Townscapes and urban sceneries: dominance of the visual

> The historic centre itself [...] is characterized by its faithfulness to the 16th century plan, the density of its monuments, and the homogeneity of its construction on a hilly and picturesque site which exalts the urban scenery by providing plunging and ascending views of incomparable beauty.[29]

This excerpt, articulating the outstanding universal value of Salvador de Bahia, brings together the monumental value, discussed above, as well as aspects of the picturesque, homogeneous and visual, which will be the focus of the present section. Privileging visual perception and seeing has a long history in Western societies, especially beginning with the late eighteenth century. Sight is viewed as the most reliable of the senses, and its role has been emphasized as part of scientific legitimacy, religious and other symbolisms, optical entertainment, aesthetic appreciation, tourism and the "culture of display."[30] There also exists a long history of privileging visual culture in the European tradition of heritage valuation following the tradition of art history and landscape studies,[31] although aesthetic considerations have not been entirely foreign to East Asian art philosophies either.[32] It is visual culture that often lends the material objects of heritage "the means of representation and achievement of meaning."[33] Against this background, it is hardly surprising to find that aesthetic valuation and conventional statements of beauty have also featured prominently as part of the definition of the outstanding universal value, or that ugliness has not played any part in the projected image. Artistic value, to be understood as a combination of aesthetic and historical values, is explicitly mentioned as one of the few values in the World Heritage Convention's definition of heritage. Under scrutiny in this section are the specific ways in which aesthetic value has been understood and articulated in relation to World Heritage inscribed cities.

The most direct references to beauty and picturesque qualities as part of the statements of outstanding universal value may be found in the ICOMOS evaluations written before the late 1990s, and, like the statements of monumentality, often deal with Central and Southern European cities. The aesthetic qualities and beauty of Segovia, San Gimignano, Kutná Hora and Salzburg, along with many other cities, were assessed to be the highest possible.[34] Outside the European realm there "are few cities in the world which are as rich as Cairo in old buildings, filled with historic significance and formal beauty,"[35] or no other such a "fine example of artistic and pictorial quality" as Sana'a.[36] These kinds of statements become almost non-existent in the twenty-first century, reflecting the evolvement of cultural heritage criterion i. In its negative statement concerning the nomination of Sibiu ICOMOS accordingly stated that being an "attractive place" is not enough to allow inclusion on the World Heritage List, if the outstanding universal value of the city has not been otherwise established.[37]

Despite the disappearance of the most direct statements regarding the picturesque value and beauty of architecture, it may nevertheless be argued that the more recent ICOMOS evaluations also treated cities as a primarily visual category. Great importance has continuously been placed on the question of how cities appear on the surface. Thus, to give one example, Goiás was considered to be "a good example of the *appearance* of a mining town" of the eighteenth and nineteenth centuries.[38] Together with monumentality, beauty has continued to operate as a rhetorical guarantee for outstanding universal value. I will now raise three points to further support this argument.

First, inherent to the ICOMOS' articulation of the outstanding universal value of cities is an assumption that an understanding of a city comes through understanding its townscape features. Much of the discussion about World Heritage cities over the years has involved form, scale, height, building volumes, style, facades, color or decorative elements.[39] A World Heritage city has mainly been understood as building blocks, or occasionally as empty plots. There has also been a strong preoccupation with typology, and ICOMOS often contemplates cities and towns from a distance as urban sceneries and skylines. This is a common language of architecture and of heritage valuation, an experts' language that objectifies the environment and observes it from a distance and at the level of abstraction.[40] For inhabitants, however, the environment constitutes a part of self-identity, a place, and a home.[41] Julian Smith summarizes this two-, or actually three-layered, beauty of an historic urban landscape. The first layer is the beauty of the landscape that we look at. The second is that of the landscape that we travel through. The third, "and deepest form of beauty, is that of the landscape that we inhabit." Smith criticizes the Vienna Memorandum, which according to him, despite hinting at "a more explicit recognition of the dynamic condition of economic and social realities, and the emerging ideas of landscape, did not diverge very far from a modernist world view."[42] Moreover, contemporary architecture in historic environments, the main concern of the Vienna Memorandum, is considered on the basis of this worldview as "inserting new visual objects into the existing, large, complex, visual object that is the historic urban landscape."[43] A very similar observation can be presented with regard to the language used by ICOMOS when defining the outstanding universal value of cities.

A second point that needs to be raised concerning World Heritage cities as a visual category is that there has been a recurrent, and, one might add, a dominant application of terms such as "harmonious," "unity" and "homogeneity" when discussing the outstanding universal value of cities, and, in particular, their townscape value. In the evaluation texts, a town as a harmonious ensemble "sui generis" has been a recurrent value.[44] Another frequent theme of what can be considered 'harmonious' has been related to the harmonious relationship of the built fabric of a historic city to the surrounding natural or rural landscape.[45] Harmonious could also represent "the harmonious fusion of different cultural traditions to produce an urban landscape of outstanding universal quality," as in the case of Lijiang.[46] On the other hand, some form of "disturbance" with regard to harmony has often been presented as an antithesis to the outstanding universal value. For Valletta, luckily, the improvements of the eighteenth century "have not disturbed this harmony."[47] In some cases, an assurance was given by the national and local authorities that they would demolish some buildings "not considered harmonious with the character of the landscape."[48] After confirmation by the Polish authorities of a future policy of demolishing some "unsympathetic recent structures in the immediate vicinity of the proposed World Heritage Monument," there were no obstacles to the inscription of Zamosc, in 1992, on the World Heritage List.[49] This suggests that it has been important for ICOMOS not only to preserve the appearance of an urban landscape, but to correct it in the direction of a more harmonious state. An implicit intertextuality and dialogue with the Venice Charter, "embodying familiar principles of heritage management without explicitly stating them,"[50] can be found here: the Charter's principle of revealing the underlying layers of the building if what was removed was of minor interest was adjusted to fit the context of World Heritage cities.

Without any doubt, two of the most central attributes considered in relation to World Heritage cities over the entire period of research were townscape unity and homogeneity. These concepts, like the general emphasis on the visual, has a long tradition as part of urban conservation initiatives, even "to the point of pastiche and replication of existing vernacular details."[51] Recent decades, however, have seen growing criticism among heritage professionals regarding the creation of homogenous cityscapes, since this activity has been seen as motivated by falsification of history and postmodernist historicism.[52] This often occurs in the interests of the tourism industry, as tourists, on average, yearn for harmony, and not for diversity.[53] Even though pastiche has many times been openly denounced by ICOMOS, the quest for unity is evident when looking at the ICOMOS evaluations of cities and the statements of outstanding universal value.[54] In the evaluation texts one finds surprisingly little of the conservationists' general concern with respect to the creation of homogenous cityscapes. The purpose here is not to dispute in any way the townscape unity that characterizes many cities on the World Heritage List, even though a well-argued case could be made that all historic cities, to a varying degree, contain elements of both unity and diversity. Instead, my intention is to point out that unity and homogeneity have been deliberately and overtly stressed by states when writing nomination dossiers, and by ICOMOS while evaluating them. Furthermore, these concepts have not been subjected to any critical scrutiny, and very little similar emphasis on townscape diversity may be found, even though it might sometimes have been clearly justified.

In the ICOMOS discourse, townscape unity has usually been equated with a well-preserved environment, whereas diversity and heterogeneity have often signaled ill-treatment or a threat to heritage. Thus, it was supposed that in Valparaíso, where spontaneity of construction has resulted in a relatively heterogeneous ensemble, "the city has many problems to solve in relation to conservation and planning."[55] A comparative analysis by ICOMOS considering Bordeaux vis-à-vis Naples suggests that when reviewed comparatively, architectural unity has been favored over diversity.[56] And even though ICOMOS did state townscape diversity to be a key attribute for World Heritage value for a few cities, even in these cases, the successive phases were often emphasized in order to underline unity as an end product:

> The historic buildings [in Vilnius] are in Gothic, Renaissance, Baroque, and Classical style (with some later additions). [...] They constitute a townscape of great diversity and yet at the same time one in which there is an overarching harmony.[57]
>
> Along a typical road there are rows of narrow shophouses or townhouses (usually of two stories) and religious buildings of different faiths. Each one is different, but with an overall sense of unity. [...] In the nominated area of Melaka, more than 600 shophouses and town houses exhibit diverse styles and influences, from Dutch style to early modern.[58]

The emphasis on townscape unity as a characteristic feature for World Heritage cities has continued to play such an important role throughout the implementation of the Convention that it has become an approach difficult to escape. Rather illustratively, the World Heritage Cities Programme specialist, Ron van Oers, while describing the innovative character of the World Heritage inscription of Valparaíso in 2003 and the random processes and dynamics related to its creation, still ended up highlighting the remarkability "that over centuries an urban landscape developed, with a vernacular architecture covering some forty-three hills, that is very homogeneous in its use of modest building schemes and materials related to the industrial era."[59]

Here the definition of the groups of buildings category for cultural heritage stated in Article 1 of the World Heritage Convention has been decisive and, one might add, also restrictive in terms of the representation of cities within the World Heritage framework: homogeneity is presented as one of the preeminent qualities determining the outstanding universal value of groups of buildings.[60] For the category of sites, or the subsequently introduced concept of cultural landscape, similar homogeneity is not expected and, overall, these categories offer more flexibility. With a few exceptions, however, all World Heritage cities discussed within the framework of this research have been inscribed on the World Heritage List as groups of buildings, even the most recent ones, although neither the Convention text nor the Operational Guidelines deny the possibility of also applying the category of sites in reference to cities. In addition to those cities officially considered as cultural landscapes, these few exceptions of cities labeled "sites" include Karlskrona, Vienna and Assisi.

In light of the above discussion the association of cities with the category of groups of buildings seems highly outdated. In fact, one of the conclusions of an expert meeting held in 2007 was that the Operational Guidelines should be revised by including the category of 'sites' as an additional category for nomination of historic cities in order to facilitate a more holistic approach to heritage.[61] A wider debate was initiated in relation to the adoption of the UNESCO Recommendation on the Historic Urban Landscape in 2011. The expert meeting which convened in Rio de Janeiro in 2009 recommended major changes to the Operational Guidelines in order to include the notion of Historic Urban Landscape in its relevant sections. The meeting proposed a change in terminology from 'historic towns' to 'urban areas and settlements,' which it considered more inclusive.[62] It also proposed a major revision to Annex 3 in the Operational Guidelines, which discusses the nomination of historic cities to the World Heritage List. The definition of inhabited urban areas and settlements that the expert group recommended departed from the visual and material understanding of the city, as it emphasized that World Heritage cities should be treated as "living entities that need to be viewed in their entirety as complex formal and spatial organizations with multiple social, economic, cultural and environmental processes that may also include a religious or symbolic component, as well as a relationship and associations with the natural environment."[63] Neither of the above-mentioned revisions have yet been made; the urban heritage guidelines added to the Operational Guidelines in 1987 have remained unaltered in the later versions of the document.[64] This has left urban heritage surprisingly untouched by the recent debates concerning the conceptual broadening of World Heritage.

The final and third point in support of the argument that a World Heritage city has been understood as a principally visual category relates to the concept of integrity, officially introduced in 2005 as a qualifying condition for cultural heritage – formerly considered for natural heritage only – but already discussed by ICOMOS on a regular basis in the context of urban nominations from 2001 onwards.

The Operational Guidelines define integrity as "a measure of the wholeness and intactness" of heritage and of its attributes. This should be approached by assessing whether the nominated area is of adequate size and includes all elements necessary to be able to convey its outstanding universal value, and by judging whether it has suffered from harmful developments.[65] ICOMOS elaborates on these issues meticulously in its evaluations. These considerations, however, often transmit an understanding of integrity as townscape, architectural and material integrity only.[66] Another side of the same coin is that ICOMOS mainly perceives the threats to cities' integrity in the form of tall, high-rise buildings, or in the form of otherwise visually unsatisfactory later construction "disturbing the appropriate perception of the property."[67] This interpretation of integrity may not be directly derived from the Operational Guidelines, which also state that the "[r]elationships and dynamic functions present in cultural landscapes, historic towns or other living properties essential to their distinctive character should also be maintained."[68] This notion suggests that socio-cultural integrity should also be a relevant consideration along with structural and visual forms of integrity.[69] While there are some recent considerations pointing more towards this direction, such as ICOMOS' notion that the urban fabric of Bridgetown "has not been shown to reflect a coherence that in turn relates to its function and history,"[70] overall this aspect remains to be systematically explored.

World Heritage cities in their territorial relationships

The Global Strategy called for considering each heritage site in "the multiple reciprocal relationships that it has with its physical and non-physical environment."[71] According to Ron van Oers precisely this kind of contextual understanding has come about for the category of cities from the broadening of meaning of World Heritage in the new millennium.[72] Indeed, cultural landscapes both formed around one city and including several cities have been inscribed on the World Heritage List in the post-Global Strategy period. At the same time, the majority of urban inscriptions, also after the launching of the Global Strategy, have been traditional historic city centers. This was, perhaps, long due to the States Parties' reluctance to use the nomination category of cultural landscape because of its supposed difficulties.[73] The Argentinian nomination of Buenos Aires as a World Heritage cultural landscape (covering 3,280 hectares) and the successive negative assessment by ICOMOS show that ICOMOS, too, has been hesitant to create new World Heritage urban landscapes. The organization recommended a possible new nomination of a more narrow area from the historic city center as a group of buildings.[74]

Knowing this emphasis on traditional historic cities, it seems most relevant to ask how ICOMOS in these instances has articulated the contextual relationship. Here it should be pointed out that much of ICOMOS' and the World Heritage Committee's concern for 'setting' has concentrated on buffer zones. The purpose of buffer zones in the World Heritage system is to protect the designated sites from various negative influences originating from their immediate surroundings.[75]

Even though buffer zones themselves are not considered to be of outstanding universal value, managing their use and development enables the value of the designated area to be maintained. During the early implementation of the Convention both the borders of the World Heritage areas and of their buffer zones were often defined in an artificial and random manner: buffer zones were either left undefined, were drawn very narrowly, or were defined but left unspecified with regard to their legal implications.[76] The more systematic definition of buffer zones and the treatment of them as operational conservation tools within the World Heritage framework is relatively recent, relating, on the one hand, to the introduction of the concept of integrity in the management of cultural heritage, and, on the other hand, to the perceived accelerated threats to World Heritage areas from their surroundings. In many ways, Vienna represented a precedent concerning the operational use of buffer zones as part of the World Heritage system. This recent interest in buffer zones and delimitation issues indicates a growing concern on the part of the World Heritage community with the contexts of World Heritage sites. The question that nevertheless remains is what type of concern this has been.

Before discussing the post-Global Strategy period, one should briefly look at the articulation of the relationship of cities to their surroundings in the early stage of ICOMOS evaluations. It can be concluded that by no means did ICOMOS leave this relationship entirely unarticulated. Instead, some cities were viewed in connection with their environmental relationship, and this relationship was approached in multiple and varying ways. The contextual view was encouraged by ICOMOS when applied by the state,[77] and sometimes criticized when not applied.[78] On a few occasions, such as in the case of San Gimignano, ICOMOS proposed that the perimeter of the World Heritage area should be extended to include parts of the surrounding rural environment.[79] It is noteworthy that there are examples of urban landscapes from the early implementation phase of the Convention (the Natural and Culturo-Historical Region of Kotor, the Ohrid Region, and M'Zab Valley), even though they were not denominated as such. Moreover, it becomes clear that from the very beginning ICOMOS was concerned about contemporary construction in the vicinity of urban World Heritage areas, especially the height, scale and visual impact of these constructions on their surroundings.[80]

Some recent ICOMOS evaluations articulate the contextual relationship at more length and in more depth. One example is Assisi, the landscape relationship of which for ICOMOS is articulated by the ancient and medieval road systems, by "the fundamental connection between the town and the valley," and by the continuous use of agricultural land.[81] Bordeaux provides another example of ICOMOS' recent understanding of the environmental relationship in its references concerning the historical and present relationship between the city and its surrounding wine-producing region.[82] Furthermore, in some of the evaluations concerning port cities, the relationship with the sea has become an actively discussed environmental relationship on the part of ICOMOS.[83] Finally, an environmental relationship was innovatively articulated in reference to the Ethiopian city Harar Jugol as a physical but also a cultural relationship:

> Harar Jugol with its surrounding landscape is an outstanding example of a traditional human settlement, representative of cultural interaction with the environment. The social and spatial structure (afocha) and the language of the people all reflect a particular and even unique relationship that there developed with the environment. The cultural and physical relationships with the territory have survived until today, but they are also vulnerable to irreversible change under the impact of the modern globalizing world.[84]

Based on the above examples, the argument that a more contextual view of World Heritage cities has developed in recent years seems well warranted. Nevertheless, when viewing the entire body of ICOMOS evaluations, it may be discerned that the articulation of the environmental relationship of a World Heritage city has been subject to diverse practices. For one thing, it is even possible to find several recent ICOMOS evaluations in which the city–environment relationship is discussed only in a very technical manner in reference to the concepts of integrity and buffer zones. Moreover, as suggested in the previous section, the environmental relationship of a World Heritage city has remained powerfully determined by visual contemplation, and, conversely, to a much lesser degree by historical, cultural, social or economic aspects.

Furthermore, in the period following the adoption of the Global Strategy the environmental relationship has continued to be treated almost solely as a traditional urban–rural relationship. The ideal setting for a World Heritage city is a rural environment having high landscape values. Actually, much of the description of the environmental relationship of Assisi dealt with the present stability of the conservation situation in the valley surrounding the city. Another variation of an ideal setting has been an isolated, hard-to-approach mountainous landscape, as exemplified by San Marino and Mount Titano.[85] Consequently, the ideal environmental relationship shows itself to be an unambiguous relationship, like that for Assisi and Harar Jugol, even though in both cases ICOMOS also identified social and economic changes that could potentially threaten the high environmental integrity in the future.

Whether the lack of this kind of ideal environmental relationship has had consequences in terms of actual World Heritage inscription has varied. On the one hand, for Úbeda and Baeza, considered to be suffering from "unattractive structures and roads" in contrast to the generally pleasing landscape of the region, ICOMOS rather laconically recorded the absence of the ideal setting but was nevertheless positive in relation to their World Heritage inscription.[86] On the other hand, the distorted relationship with a setting that used to be rural was one of the key reasons why ICOMOS was not in favor of the nomination of the Argentinian city of La Plata in 2007. ICOMOS was concerned about many aspects of the nomination; however, the fact that there had been more recent suburbanization and that the grid plan had been continued into the surrounding rural landscape "thus obfuscating the strict distinction of the urban and rural areas," was considered a major flaw.[87]

The notion of accelerating out-of-scale and out-of-place projects in urban World Heritage contexts has continued to be decisive in the ways that ICOMOS has articulated the environmental relationship of a World Heritage city. Discussion concerning the contextual relationship has often been concerned with the threats posed by urban surroundings to the historic city – its negative potential for high-rise constructions and for producing more urbanization and suburbanization. Therefore, it seems that conceptually very little variation has existed between those cities harmoniously blending with rural settings and those cities troubled by their urbanized or urbanizing surroundings.

Here the implementation of the World Heritage Convention has not departed greatly from the Venice Charter definition of the "setting," according to which monument and setting should be clearly separated and "wherever the traditional setting exists, it must be kept."[88] Particularly, the defining of buffer zones has sometimes meant establishing hierarchies and creating artificial boundaries between the area of World Heritage value, and a wider urban context considered less significant. Determining outstanding universal value also has much to do with delineation. A proper determining of boundaries is important to ensure the effective protection of a site, but it also involves making a selection and is susceptible to political negotiation.[89] The difficulty of this exercise is made explicit by the example of Tel Aviv, where the buffer zone border cuts through one of the city's main streets. This has led to the peculiar question of how to treat, planning-wise, the different sides of the street – one included in the buffer zone and the other not.[90] Recent ICOMOS and World Heritage Centre publications, however, challenge this long-held view by highlighting the role of the setting as an "equal, complementary and inseparable part of the so-called zone of primary importance," or even as an integral part of the inscribed zone.[91] These novel openings emerge in the most recent ICOMOS evaluation texts, even though they too define the environmental relationship as mainly a physical and visual relationship. If elaborated on more in the future, they have the potential of encouraging a more holistic understanding of a World Heritage city. What is obvious is that the environmental relationships of historical cities and the various influences on their outstanding universal value reach well beyond the boundaries of the designated areas or buffer zones.

Outstanding universal value and urban life

The previous sections have mainly discussed the ways in which the dominant historical, aesthetic and architectural values have been articulated in the context of World Heritage cities. The purpose of this section is to examine the role and integration of social and local community values within the ICOMOS discourse on World Heritage cities. The focus will be on issues such as everyday heritage, references to urban life, citizens and daily comings and goings in the city, the function of heritage in the community, and local community involvement. Over the following pages I also discuss the articulation of the relationship between intangible and material heritage components in the definition of outstanding universal value.

ICOMOS and social values

Jukka Jokilehto, while distinguishing between the World Heritage List and the UNESCO list of intangible heritage, argues that while focusing on a living historic town, such as Marrakesh, the World Heritage List, on the one hand, "would certainly recognize that life goes on in the town and that this life and social functions are essential elements in the definition of universal value of the place." On the other hand, the list of intangible heritage focuses "on activities and processes that have traditionally been and continue taking place in a specified cultural space of the town, the principal market place of Marrakesh."[92] Here Jokilehto's intention is to point out both the critical differences between the two parallel UNESCO Conventions and their ability to complement each other. What is interesting in connection with our discussion is that Jokilehto chooses to present Marrakesh as an example of a World Heritage city where "life goes on," since Marrakesh is one of the rare examples among World Heritage cities in connection with which the present city and the life it contains have been given such an openly positive connotation as part of the definition of outstanding universal value:

> With its maze of narrow streets, its houses, its souks [markets], fonduks, its traditional crafts and trade activities, its medina of 700 hectares, this ancient habitat, which has become vulnerable due to population growth, constitutes *an eminent example of a vibrant historic city.*[93]

In the early period of the implementation of the Convention, when Marrakesh was also included in the World Heritage List, ICOMOS mostly stuck to an architectural understanding of place and refused to include social and cultural values as part of its definitions of outstanding universal value. Djenné, thus, provides a more common example. The outstanding universal value of Djenné had been described in the Malian nomination documentation, along with more traditional attributes, as the result of the position of the city as a vibrant social, cultural and religious center.[94] The subsequent commentary by ICOMOS in 1981 touched upon the living city from a rather different angle, with a brief reference to a "group of living quarters" continuing to "encroach" upon the large public square close to the mosque.[95] As such, this was obviously a justifiable concern in light of the uncontrolled urbanization taking place in Djenné. The second evaluation document from 1988, following a new nomination by the Malian Government, established the outstanding universal value of Djenné based mainly on architectural features and excluded any reference to the social organization of the city: "Djenné is an outstanding example of an architectural group of buildings illustrating a significant historic period. It has been defined both as 'the most beautiful city of Africa' and as 'the typical African city.'"[96] The locus of value was relocated away from social dimensions.[97] Here we should bear in mind that whilst the Australian Burra Charter mentioned social value in 1979 as one of the four attributes for cultural significance, the implementation of the World Heritage Convention, in the same period, took a different path. The first criteria of 1977 had still included references to social dimensions of heritage, but these were soon removed, except for the notion of a traditional human settlement in criterion v. These fine arts–focused criteria did not come as a given to ICOMOS. The organization played an active part in establishing them, and has been working within their parameters ever since.

In the new millennium, there has been a tendency to include a wider array of themes and values in the definition of World Heritage cities, something that may be perceived as a direct outcome of the adoption of the Global Strategy, the Nara Document on Authenticity and the 2003 Intangible Heritage Convention. It was, for example, noted that Goiás "has an important meaning for the local community, not only on account of its urban and architectural values but also for its rich social and cultural life,"[98] and that in Vienna "the historic town is conceived not as a museum but rather as a living and vibrant city."[99] In reference to Verona, ICOMOS elaborated on the role of the new urban management plan in rebalancing the social and economic structure of the historic town. However, this kind of concern with gentrification of historic cities has otherwise been almost non-existent in the ICOMOS valuation discourse.[100] Multiculturalism reflected in the present-day urban community, as in the case of Macao and Melaka and George Town, presents a theme in the context of which ICOMOS most often has chosen to discuss continuity in terms of social community.[101] The ICOMOS evaluation on Macao provides an example where cultural exchange between Europe and the rest of the world, in this case China, was understood as a broader phenomenon than just architectural influence and interchange, involving even the Creole language.[102]

Despite the above examples, and a few others, the overall conclusion must still be that in the case of most cities the obverse of the emphasis on artistic and architectural values, great-power histories, and visual aesthetics has been the overall lack of discussion concerning social values. This is true for the consideration of past urban life and social and cultural urban histories, as shown in Chapter 4, and it is equally true for the consideration of current forms of urban life. For the most part, social values, locating the value of places within people rather than in fabric whose significance is determined by professionals,[103] have been excluded from the descriptions of World Heritage cities. Even several recent evaluation texts fail to establish the connection between past and present urban life altogether. For example, while ICOMOS chose to comment on the "ignorant" changes made by shopkeepers in Le Havre, it failed to discuss the role played by those shopkeepers in terms of functional continuity.[104]

What thus becomes obvious is that despite the more recent overall World Heritage rhetoric, social and intangible values continue to be treated as secondary values in the valuation process. This is visible in the World Heritage evaluation criteria, and it becomes equally evident when considering the individual statements of outstanding universal value of cities, such as the one highlighting the particularly high value of Bardejov, "because of its present-day vitality and contemporary activities which do not compromise the other values." While the present-day urban community was progressively regarded as presenting a value of its own, it was given status secondary to other values, defined as urban, architectural, historic and aesthetic.[105] A common approach to social values has been to include a sentence or two noting, for example, the existence of religious practices or festivals continuing to take place in the city. When treated in this way, and, when at the same time architectural values and tangible heritage elements are discussed at length, the position of social values as second-rank values is made explicit. Even in the case of Melaka and George Town the inclusiveness of social values and intangible heritage in the valuation of these two places was somewhat undermined by a commentary elsewhere in the document on the importance of intangible heritage, if "related to the tangible components." Also, even though intangible heritage in the context of Melaka and George Town was seen to reflect a wide array of practices, the practices related to trade were, again, not included as part of this definition. What was highlighted, instead, was the materiality and different architectural styles of shophouses.[106]

The next question that arises from the above considerations is whether or not there has been a marked difference between the articulation of outstanding universal value for European and non-European cities in terms of social and intangible values. This is especially relevant knowing that Asian and African conceptions of cultural heritage have primarily focused on intangible and spiritual heritage aspects and the interchanges between community and environment.[107] The overall answer to this question is positive – ICOMOS' reluctance to include social and intangible dimensions of urban heritage has concerned European cities more than their non-European counterparts, with very few exceptions.[108] It has been more natural for ICOMOS to describe Islamic cities as busy places of commerce, or African cities as places for continuing traditions and religious practices, than to describe European cities in the same way. In a way, ICOMOS is fulfilling here the Global Strategy's and the Nara Document's objectives by viewing each heritage item in its cultural context; the relevant cultural context for Europe has simply been understood as one focusing almost solely on material heritage aspects. But it may be also argued that ICOMOS has found altogether very little rationale for providing a wider value-basis and discussing intangible heritage aspects as regards cities, Western or non-Western, in the context of which the traditional tangible heritage aspects and conservation of their material authenticity figure strongly. One such non-Western example is provided by Saint Louis in Senegal, which, in 2000, was discussed as with any European city in terms of its physical components and architectural history.[109]

Local community involvement

In contrast to the earlier ICOMOS evaluations which often did not even mention the existence of local populations in World Heritage cities, several ICOMOS evaluations since the late 1990s have touched upon local communities. When considering the recognition of local communities as carriers of social values, it needs to be remembered that local community involvement was first added as an independent section to ICOMOS evaluations, and to national nomination documents, as late as 2007. The examination of the more recent ICOMOS evaluations, especially the ones compiled between 2007 and 2011, shows that in most cases the involvement of local communities has been understood to mean the involvement of official city governments,[110] one-sided information provided by the state/municipality about the nomination and management guidelines to the owners of the buildings,[111] or something taking place through formal citizens' associations.[112] In the view of ICOMOS, local community participation would also involve, as in the case of Ibiza, local training workshops teaching proper restoration techniques under the supervision of "experienced architects and archaeologists,"[113] or as in the case of Mantua, "the action of a real estate Commission, consisting of experts, whose mission is to assess projects that could affect the external appearance of buildings."[114] Even though the local community's attachment to the place, and the citizens' positive contribution to the preservation and conservation of the site have been mentioned,[115] in most cases, local community involvement has been understood to mean the local population's acceptance of protection and its awareness of the historic, artistic and architectural values identified by experts.[116]

Chapter 2 discussed the Operational Guidelines' references to local communities and how they are made discursively secondary to other stakeholders by referring to local communities as "partners" instead of "parties," and by using a low degree of modality, therefore implying that the knowledge of local communities is open to interpretation. Waterton, Smith and Campbell show how in the Burra Charter text the authority of experts is reinforced by various discursive legitimizing techniques, instead pushing non-expert participants "into the role of beneficiaries."[117] These kinds of legitimizing techniques are also characteristic of the descriptions of local communities in ICOMOS evaluations. The use of verbs such as "educate," "demonstrate" and "raise consciousness" is illustrative in this sense. For instance, in reference to Bridgetown and its Garrison ICOMOS noted with satisfaction how "the Barbados World Heritage Task Force [...] has developed and implemented a public awareness and education programme, which caters specifically to educating Barbadians about the nominated property."[118] On the other hand, in its negative assessment of the outstanding universal value of Jeddah, ICOMOS regretted that the 1980 Rehabilitation Plan had "not fully succeeded in raising citizens and real estate owners' consciousness of the importance of Jeddah's built heritage."[119] The whole purpose and meaning of ICOMOS statements is to give an expert *evaluation*. In this framework experts always come before non-expert communities, and experts are the ones who establish meaning and value and who, if needed, teach these historical and aesthetic values to the local population. One brief paragraph discussing the involvement of local communities can hardly alter this.

In the official World Heritage discourse local communities are – paradoxically – sometimes considered a threat to World Heritage.[120] One of the main threats that ICOMOS identifies in the urban context, development pressures, can be placed, at least partially, within the local community. It is also possible to identify examples where ICOMOS, perhaps unintentionally, has placed the protection of the visual appearance of the city in opposition to the needs of its present users, in favor of the former, as suggested by the extracts from the evaluations on Le Havre and Acre:

> The risks to the area include the changes made by the population. For example, some shopkeepers, ignorant of the regulations and wanting to make their signs more visible, have tried to enlarge them excessively, give them inappropriate supports and lighting, and allow them to trespass on the public domain.[121]
>
> The most serious problem confronting those responsible for the conservation and maintenance of the old city is a social one. [...] Few of the present-day inhabitants have any family ties with the city and so there is a lack of identification with it. Furthermore, many of the inhabitants are unemployed or poorly remunerated and so cannot afford to live elsewhere. If and when their personal fortunes change, they will immediately seek housing outside the walled city. As a result, they do not feel themselves under obligation to respect the *appearance* of what is to them no more than a transitory place of residence.[122]

ICOMOS ends its recommendation with the notion that there should be a social programme intended to improve the quality of life in Acre, thus ascending to a more social understanding of the situation.[123] What the above extract nevertheless suggests is an understanding that the problems are social, whereas the value to be protected is aesthetic.

It thus seems that even the most recent official World Heritage discourse has projected a somewhat limited understanding of what constitutes a community, or community participation. In the context of World Heritage the main consideration of community has so far been placed, in the 1970s scholarly understanding of the term, on agrarian, ethnic minority, or indigenous "face-to-face and traditional collections of people."[124] This definition has largely excluded urban communities which, as carriers of social values, have been only a very recent addition to the discourses on outstanding universal value of cities. ICOMOS took a clearly articulated interest in local community aspects as regards Harar Jugol, by describing its social organization according to neighborhood associations (*afocha*). This interest perhaps resulted from Harar Jugol's consideration in the World Heritage framework as "a traditionally functioning community."[125]

Furthermore, and analogous to the overall idea of harmonious World Heritage, in ICOMOS evaluations the idea of a local community has most often appeared as an idealized version of a community as an uncontested and unified group and a fixed entity whose more problematic dimensions have rarely been addressed.[126] Acre, referred to earlier, provides an exception to this rule, as ICOMOS, as part of its evaluation, identified the present-day Palestinian population that has felt disassociated from the heritage of the previous inhabitants. While most World Heritage cities undoubtedly form essential reference points in respect to their resident communities' identities, Acre clearly fails to do so. In Mostar, the World Heritage area bears significance for only part of the local community, as was pointed out in Chapter 4. In Le Havre, even though the city is highly appreciated by the international expert community as a representative of post-war reconstruction, the process of developing local appreciation is still very much in progress.[127] These examples show that urban community in the context of World Heritage is far from an unambiguous notion. There would seem to be a constant need for problematizing what constitutes a community in the context of World Heritage cities, and also with regard to those communities that are seemingly conflict-free.

It is also telling that the section discussing local community involvement is hierarchically placed under the title "Management" in the ICOMOS evaluation document. This implies that the local community is managed along with the rest of the heritage. In some cases it becomes obvious that ICOMOS thinks that the local heritage authorities also need managing from above.[128] Moreover, it should be pointed out that in the ICOMOS evaluation texts community involvement has been discussed, as with other references to social and intangible heritage values, in a standardized fashion. By reading these documents, it becomes obvious that the standardized nomination and evaluation formats do not support diverging views of community, but instead create rather repetitive and meaningless accounts. What is more, ICOMOS has not been in a position to 'enforce' local community representation and participation, probably because this issue has been considered as falling within the realm of national sovereignty. Concerning Yaroslavl, ICOMOS, for example, was able to note only that even though, according to present legislation, civic associations are not allowed to get involved in the decision-making process, these associations nevertheless "represent a potential for the future."[129] It thus seems that ICOMOS has sometimes engaged in embracing "the rhetoric of community," when it has found itself "in the midst of a political and social context rife with exclusion, intolerance and injustice."[130]

Physical fabric versus intangible urban heritage

Perhaps the most flexible understanding of outstanding universal value with regard to the relationship between intangible and tangible heritage aspects in the context of cities was presented by the 2009 inscription of the once-rejected urban site of Cidade Velha, the historic center of Ribeira Grande in Cape Verde. ICOMOS, in 1992, acknowledged the "indisputable" historical significance of Ribeira Grande as the first colonial city to have been established by Europeans in the tropics. The organization, however, was not favorable to Cidade Velha's inscription for the reasons that the historic center lacked sufficient integrity and authenticity and the contemporary town had almost entirely overrun the historic fabric. This negative position based on Cidade Velha lacking tangible evidence to its World Heritage quality had changed, however, by the time of a new nomination presented by Cape Verde almost two decades later. The State Party presented certain new aspects in support of the new nomination; most importantly it introduced Ribeira Grande as an early center for experimentation in the field of colonial agriculture and for the acclimatization of plants between continents – another expression of the recent tendency towards the inclusion of transnational histories as part of the narratives of outstanding universal value of cities. Moreover, according to ICOMOS, the management, authenticity and integrity of the site had improved since the first examination, and thus the earlier obstacles regarding these aspects had been removed. Still, but importantly, the whole essence of Cidade Velha's nomination and the ICOMOS statement of outstanding universal value concerned the historical significance and intangible heritage aspects of the city. As summed up by ICOMOS, the "historical value of Cidade Velha/Ribeira Grande is undeniable, as it brings together in a single place a high density of important events in human and cultural interchange over a very great distance. However, compared with this historic wealth the monumental heritage is only partial."[131] Here ICOMOS was willing to acknowledge that intangible historical associations of a city counted for more than its physical remains, an admission that was made very much in accord with the Global Strategy's objective of inclusiveness for non-European heritage. Cidade Velha, which is the only architectural ensemble listed as National Heritage of Cape Verde, became the first site designated as World Heritage from that country.

It is too early to say whether the inscription of Cidade Velha represents a new less tangibly focused trend in the definition of outstanding universal value for cities. Many other recent statements issued by ICOMOS, in contrast, suggest a continuing strong commitment by ICOMOS to material heritage in support of the outstanding universal value. This could be seen in 2008, as ICOMOS gave a recommendation of inscription on the World Heritage List to the city state of San Marino, the outstanding universal value of which was argued by the State Party mostly in relation to the city's historical associations with the idea of independent city states. As a condition for a positive statement, ICOMOS required the State Party to demonstrate the links between the non-material and material heritage aspects of San Marino.[132] ICOMOS evaluations on two rejected nominations further support this argument. Gdańsk provides the first example. The city was first proposed for inclusion on the World Heritage List in 1997 based on a relatively traditional historic city approach, as may also be read from its title "Gdańsk: The Main Town, the Motlava Side Channel, and the Vistula Mouth Fortress." According to ICOMOS, the case for the exceptional nature of Gdańsk in historical and cultural terms, as proposed in the nomination, had not been substantiated.[133] At the second attempt, Poland tried to build a case for the inscription of Gdańsk as the "Site of Memory and Freedom." The nomination included a select list of 12 historic, mostly medieval buildings which had survived the destruction which occurred at the end of the World War II. These were introduced as illustrative of Gdańsk as a Free City and as a town of religious tolerance. Furthermore, the Polish nomination included two significant memorial sites associated with later events: the Westerplatte memorial site, recalling resistance against the Nazi occupation at the beginning of World War II, and the Gdańsk Shipyard, testifying to the emergence and development of the Solidarity Movement. This clearly presented a new type of urban nomination focusing mainly on intangible heritage aspects as it combined events of historical significance from different epochs in the city under one common theme of memory and freedom. The ICOMOS verdict, however, was unequivocal. It acknowledged the importance of Gdańsk as "a symbolic city," but felt that this condition had not been sufficiently demonstrated, on the basis of material evidence, to justify inscription on the World Heritage List.[134]

The second example involves Buenos Aires, nominated in 2007 and evaluated by ICOMOS the following year. As already mentioned, the nomination concerned a large urban area of 3,280 hectares, and it was proposed by Argentina within the category of cultural landscapes. While the nomination had obvious weaknesses in terms of delimitation and the argumentation of outstanding universal value, the interest here lies in its use of criterion vi. This criterion was justified by the State Party on the grounds of "the vitality of cultural life, cultural diversity, an open-minded perspective" reflected in "bars, coffee shops, corners and clubs, literary sites or 'sites of urban memories,' as well as the use of public places, the urban scene and individual examples of architecture" – offering a clearly novel way to approach cities of the Western hemisphere in the World Heritage framework. According to ICOMOS, these values had not been justified as exceptional in the nomination documentation, and they were not "reflected in the tangible aspects of the city in a way that has a profound impact on its planning and built assets."[135] Notwithstanding the other weaknesses of the nomination of Buenos Aires, it may be asked what would be the proper way to tangibly justify the vitality of cultural life, cultural diversity, and the open-minded perspective of a city.

The debate during the 2011 the World Heritage Committee meeting, which concerned the inclusion of Historic Bridgetown and its Garrison on the World Heritage List, mainly revolved around the issue of the relationship between tangible and intangible heritage components. While ICOMOS expected more tangible evidence to support the outstanding universal value,[136] the Committee was willing to inscribe the nominated site with an intangible focus.[137] Based on the above examples and other recent debates, it seems warranted to say that while it is not yet possible to detect any straightforward change, the implementation of the World Heritage Convention has come to a crossroads regarding the relationship between tangible and intangible urban heritage values. ICOMOS tries hard to balance between how much of outstanding universal value can be founded on intangible historical significance, or on the continuity of urban forms of life, and, conversely, how much these intangible heritage components should be proven to exist in situ by the remaining physical urban fabric. In two of its recent negative statements ICOMOS utilized the concept of the "complexity of the city" to urge the States Parties to consider specific urban values of a nominated place as a complex old city. In the case of Alanya, a reflection on these urban values should have included a description of "the port city, the ancient harbor, the plan of the old town and its remains," and "the intangible values that inextricably contribute to make up the complexity of the city."[138] How these different attributes making the complexity of the city weigh in relation to each other was still left unspecified.

Social values: a category in their own right?

What Jokilehto writes about considering life and social functions as an essential element in the definition of the universal value of a place would be ideal and highly recommendable in the context of the World Heritage valuation of cities. Up until the year 2011, it may be argued, this was not the reality. That does not mean that ICOMOS was ignorant of urban life in historic cities, which it surely was not, but simply that social aspects did not figure prominently as part of the definition of outstanding universal value. Part of the failure within the World Heritage framework to discuss social and other intangible values can be related to these values being impossible to elaborate as objective and scientific in comparison to artistic and historical ones. Even though it is a generally shared view that all heritage values are socially constructed, the ICOMOS discourse has nevertheless also sustained a parallel rhetoric – one that treats historical and artistic values as intrinsic.

Even though the lack of a social dimension is perhaps in accordance with the letter of the World Heritage Convention, it is certainly not in accordance with the spirit of the later evolution of World Heritage norms – in particular the Nara Declaration's or the Global Strategy's objectives with regard to considering cultural heritage from a more anthropological perspective. Here we come back to the group of buildings category's inability to host dimensions of diversified cities, this time social dimensions. Also, it seems that the possibilities of the re-formulations of the cultural heritage criteria in the 2000s have not been fully utilized with regard to cities. At the same time it needs to be pointed out that the notions of "traditional human settlement," "human interaction with the environment" and "a cultural tradition which is living," which express cultural and social values in the current criteria, still have rural and indigenous meanings embedded in them.

The other part of the lack of discussion concerning social and other intangible values as part of the valuation of World Heritage, especially in relation to European cities, may perhaps be explained by the inner debates within the ICOMOS concerning the fundamental principles and ideology of the organization. This difference of views became articulated in two successive statements by two former presidents of the ICOMOS International: Gustavo Araoz (2008–2014) and Michael Petzet (1999–2008). According to Araoz, in 2009, "even in the Western world, the values of traditional heritage no longer reside exclusively on its physical fabric and form, but on intangible concepts that by their very nature [are] in constant flux. [...] The dispersal of values between material and intangible vessels increasingly comes at the expense of the historic fabric of the place."[139] Petzet, in his response, saw that with Araoz's writing "the core ideology" of ICOMOS, the conservation of monuments and sites, was being counteracted. According to Petzet, "respect for the special traditions of all world regions applies also to the great European tradition of conservation, which should not be discriminated against on the basis of certain 'old Europe' attitudes, which should not be misunderstood in a virtually insulting way (Europe – 200 years of 'focus on materiality,' etc.) by playing off 'tangible' against 'intangible' vessels, which are in fact two sides of the same coin."[140] This is a recent debate, but a similar kind of concern for the future role of traditional European heritage values within the discourses of ICOMOS and UNESCO may also be read between the lines of some of the commentaries concerning the Nara Document on Authenticity.[141] The question is, it seems, should the social dimensions of heritage remain the 'other' in the World Heritage valuation matrix, identified only when the significance of the place is not primarily material and aesthetic, or should they be accepted as a category in their own right?

World Heritage cities, authenticity and change: an ambiguous relationship

Out of the many meanings given to the word 'authenticity,'[142] two – genuineness, as opposed to copied, and original state, as opposed to change – have been dominant in the Western framework of heritage, including the World Heritage Convention. In actual practice, however, authenticity is often found somewhere between the two extremes. Christina Cameron asserted already in 1994 that "if strictly applied, few cultural properties would meet the authenticity test for inscription on the World Heritage List."[143] The relationship with regard to authenticity remained ambiguous even after the Nara Conference on Authenticity in 1994, as pointed out in Chapter 3. Hence, it will suffice here to look into this ambiguity further in the case of cities and as part of the ICOMOS' articulation of outstanding universal value, and by using the Nara proceedings as a dividing watershed. This section of my work thus explores first, how ICOMOS has balanced between the different degrees of authenticity (design, material, workmanship and setting, and, after Nara, also function and spirit), and second, how ICOMOS, going beyond the Warsaw example, has dealt with the reconstructed realities of World Heritage cities. The section ends in a discussion of the discursive roles given to permanence and change as part of the valuation of cities as World Heritage. It will therefore develop further the discussion initiated in Chapter 4, which argued that ICOMOS' articulation of outstanding universal value of cities has been marked by an understanding of continuity as something belonging to the safely distant past.

Different degrees of authenticity

In World Heritage practice, the condition of authenticity proved difficult to interpret and problematic to apply, both for the States Parties and ICOMOS alike.[144] In its early evaluations ICOMOS rarely referred to authenticity explicitly. In addition, the four attributes defined for the test of authenticity – design, material, setting, and workmanship – were hardly ever specified or analytically separated. Some of the discussion concerning 'setting,' at least implicitly, referred to authenticity, even though, as Henry Cleere recalled, the authenticity of the setting was the most difficult aspect to evaluate for ICOMOS.[145] The guidelines laying out the nomination process for historic towns on the World Heritage List since 1987 offered somewhat more flexibility for the consideration of the authenticity of cities: the condition of authenticity was met when the "spatial organization, structure, materials, forms, and where possible functions of a group of buildings [...] essentially reflect the civilization or succession of civilizations which have prompted the nomination of the property."[146] This flexibility notwithstanding, the authenticity of cities was also treated primarily as a material condition.

An implicit notion concerning authenticity in early ICOMOS evaluations was that material authenticity was primarily a feature of cities built of stone.[147] At the same time, and showing again a more diversified picture, there were a few instances when the continuous replacement of the building material was considered to be in conformity with the notion of material authenticity. Ghadamès was noted to have conserved its original materials: pise or clay brick walls, woodwork, masonry and palm wood casings. Nowhere in the document was it suggested that these materials had never been replaced as part of a continuous process of maintenance.[148] The same thing becomes obvious in reference to Nordic wooden towns, in particular Bryggen and Rauma, where the continuous replacement of the wooden building material was not considered as contributing to any loss of authenticity.[149] According to Léon Pressouyre, the World Heritage Committee's attitude was not so favorable concerning the inscription of constructions predominantly in earth.[150] However, the ICOMOS evaluation in 1993 of Coro, constituting one of the most important towns with buildings of earthen construction in the Caribbean region, saw no questioning of the endurance of its construction material.[151] While the fact that cities built out of materials other than stone represent a clear minority among the World Heritage–listed cities indicates that priority was given to the more permanent stone structures, the above examples nevertheless illustrate that the replacement of original material was accepted when carried out in accordance with the original appearance and original form, and when using traditional techniques.

The most traditional post-Nara ICOMOS' reflections on authenticity concerned traditional Central and Southern European historic city centers, for which the conditions of material and formal authenticity apply very well.[152] In these cases, ICOMOS saw very little incentive to consider any alternative ways of approaching the concept, and the defining of authenticity continued to be a reflection of the application of modern restoration principles, taking the Venice Charter as the guiding authority. The inscriptions of the twentieth-century cities, otherwise groundbreaking in the World Heritage framework, also allowed authenticity to be understood as an original material testimony and design, and as an unchanged status. An illustrative recent example in this respect is Le Havre.[153]

A few post-Nara ICOMOS statements refer to the Nara Document on Authenticity: out of the fifteen evaluation documents concerning urban nominations inscribed between 2006 and 2011 two, those of San Marino and San Miguel,[154] make explicit references to Nara. This lack of references is somewhat surprising, because the Nara Document was developed specifically in regard to questions related to World Heritage, and because the 2005 revised Operational Guidelines finally included the mention of Nara principles. As Waterton, Smith and Campbell point out, "the degree to which a text enters into dialogicality is expressive of a willingness to negotiate and interact."[155] The lack of references to the Nara Document suggests that this document is still treated as an external voice in the World Heritage valuation discourse, in relation to which the negotiation of difference has not been fully opened. The evaluation documents continue to interact mainly with the Venice Charter.

We may, however, find several interpretations of the concept of authenticity in the post-1994 period that are more open. In general terms, what has ensued has been a discussion about a broader range of authenticities: the authenticity of continuing building traditions, the authenticity of character, or the authenticity related to organic growth and continuous settlement. Oporto, inscribed shortly after Nara, serves as an example: "The authenticity of the urban fabric of Oporto is absolute, since it is a reflection of over a thousand years of continuous settlement, with successive interventions each leaving their imprints."[156] The authenticity of uninterrupted use and function was also emphasized more frequently after Nara. For instance, Alcalá de Henares was commended for the unusual recovery of its authenticity of function through the return of the university to the city after a century and a half,[157] and Verona for the continuity of its military use (which, however, ended soon after the World Heritage designation).[158] ICOMOS has not, however, commented systematically on functional authenticity in comparison to more traditional aspects of authenticity, which shows that the relationship between different authenticities has been unevenly articulated. The evaluation document concerning the Canal Area of Amsterdam serves as a good example. Even though ICOMOS raised the issue of the altered use of buildings (for example warehouses being converted into residences and offices) as an aspect of the consideration of authenticity, this was not followed up on in any way, and the consideration of change in relation to authenticity ultimately became an issue of material and visual authenticity.[159]

A very flexible interpretation of the concept of authenticity in the post-Nara period was formulated in relation to Campeche. Here authenticity was considered to exist "due to the continuity of a traditional family lifestyle, with manifestations of a rich intangible heritage, illustrated by local music, dances, cooking, crafts, and clothes."[160] The combined statement of authenticity and integrity concerning Valparaíso, focusing on continuity in responses to the cultural landscape, in the overall character deriving from the economic boom of the end of the nineteenth century, and in land use and construction techniques, is no doubt among the most flexible ones so far issued by ICOMOS.[161]

Such evolution notwithstanding, even the most flexible interpretations of authenticity come back to the importance of the high level of material permanency. For instance, for Melaka and George Town, the socio-cultural aspect of authenticity were hierarchically considered secondary to the more traditional aspects.[162] This was the case even though the State Party of Malaysia had chosen to describe the authenticity of these two cities in the nomination dossier beginning with "the Authenticity of the Living Heritage."[163] Even in the above-cited case of Campeche, a full reading of the statement of its authenticity reveals that the continuity of a lifestyle and manifestations of intangible heritage served as a pre-condition to "a high degree of authenticity because of the small number of transformations and interventions," in other words for the authenticity of the physical city and less for their own right.[164] The articulation of authenticity goes hand in hand with the intangible-values debate. Thus, with very few exceptions, ICOMOS has refused to define authenticity as primarily socio-cultural, functional and spiritual authenticity.

Reconstruction beyond Warsaw

The Venice Charter language has been amalgamated as part of the ICOMOS evaluations. In a few instances a selective demolition of existing structures considered to have little intrinsic value was encouraged by ICOMOS, as noted in Chapter 4. The Venice Charter also condemned reconstruction as unfitting modern conservation. ICOMOS' reversal of its authenticity statement concerning Warsaw already suggested that the organization was not going to treat the issue of reconstruction as a black and white situation. Even though in most cases reconstruction was seen as negative, there were also occasions when it was accepted. The attitudes towards reconstructions could vary, even on a case-specific basis, as shown by the three examples of Popayán, Carcassonne and Rhodes. Seventy percent of the city of Popayán had been destroyed in an earthquake in 1983; a few years later ICOMOS considered the extensive recent reconstruction detrimental to the authenticity of the city, and the nomination was rejected.[165] Despite the renewal of the appearance of the former city center of Popoyán, the replicas were constructed of reinforced concrete rather that adobe and interiors were altered.[166] In the well-known case of Carcassone, Viollet-le-Duc's nineteenth-century restorations provoked deferral of this nomination in 1985, despite the views expressed by the chairman of ICOMOS, Michel Parent, two years earlier in favor of accepting the nineteenth-century reconstructions as part of the history of the nominated site (see Chapter 3). Finally, in the case of Rhodes, despite very critical remarks concerning twentieth-century pastiches and reconstructions completed during the period of the Italian occupation (1912–1948) inside and outside the medieval city, ICOMOS was in favor of including this city on the World Heritage List in 1988. The "pseudo-medieval monuments," as the reconstructions were called, were considered a "permanent integral part of the urban history of Rhodes." At the time of the nomination, a project to rebuild a fifteenth-century tower that had collapsed in 1863 was under consideration in the city, but even this did not constitute an impediment to the inscription.[167] A similar lack of discussion regarding the principles of reconstruction was also notable in the case of Warsaw; at the moment of its inscription there was no reflection on these principles, which had been essentially based on the idea of a return to the eighteenth-century cityscape. As part of Warsaw's reconstruction, paradoxically, even a building in the Gothic Revival Style, one of the rare ones to have survived the bombing, was demolished as unfitting within the framework of the reconstruction schemes.[168] These early stage elaborations on the reconstruction of cities by ICOMOS confirm the view that the use of the authenticity condition varied significantly. The attitude also varied regardless of whether the facsimile reconstruction had been applied to post-conflict situations where reconstruction has been seen as an integral component of the general rebuilding of post-conflict societies and contributing towards reconciliation, thus often more easily permitted by the international conservation community, or to destruction by forces of nature.[169]

The reorientation of the mid-1990s was expressed by the inscription of the once-deferred nomination of Carcassonne to the World Heritage List in 1997. Eugène-Emmanuel Viollet-le-Duc's restoration works, in 1985 still regarded as unacceptable in the light of modern conservation principles, were accepted as a result of the more context-based view of authenticity outlined in the Nara Document on Authenticity. In the case of Carcassonne, the first context to be considered was that of "historic stratification, which can be verified genuine." The second context was that of the restoration works, now reviewed as a positive feature.[170] Similarly, the nineteenth-century restorations in Siena by architect Giuseppe Partini were considered in light of "renewal of arts and traditions" and as fitting well "into the continuity of the 'Gothic Dream' that Siena represents."[171] In the case of the non-inscribed Hungarian city of Pecs the degree of authenticity was considered to be high despite the existence of "a considerable amount of pastiche [...] carried out in an attempt (largely successful) to reproduce what had disappeared during the 150 years of Ottoman occupation and destruction."[172] With regard to Verona the justification for the post-war reconstruction, made necessary because forty percent of the building stock had been destroyed, was not disputed in any way.[173]

While accepting the extensive restoration and reconstruction carried out in the late nineteenth century and between the two world wars in San Marino as part of the historical evolution of the city, ICOMOS was nevertheless very critical towards the more recent restoration in the city, which in some cases had continued the policy of reconstruction.[174] Also, in 1998 the organization denied Gdańsk's authenticity because of the post-war reconstruction, which, in a very similar manner to that of Warsaw, had been founded on conscious recourse to the pre-eighteenth-century appearance of the city. While the State Party and the ICOMOS on-site mission considered this reconstruction to be in conformity with the Nara Declaration's spirit, the ICOMOS World Heritage board found this debatable because "the conscious recourse to pre-19th century appearance seems to be at odds with that document."[175] In both San Marino's and Gdańsk's cases ICOMOS drew the line between acceptable and unacceptable reconstructions at the confines of World War II. Mostar's inclusion in the World Heritage List points in again another direction: acceptance of a recent facsimile reconstruction, if connected with reconstruction of the intangible dimensions.[176]

Between permanence and change

One of the rare early explicit references that ICOMOS made to authenticity was in relation to the cities of the Gulf of Kotor. According to ICOMOS, it was the "exceptional authenticity of their conservation" that warranted consideration of these cities as unique. Conversely, the decision to exclude the Gulf of Tivat from the nominated area was considered justified "by the lesser authenticity of these cities, more disturbed by recent industrialization."[177] The Kotor example reveals that an antithesis to authenticity was considered to be recent change. In its ideal form the authenticity of cities was thus perceived of as an unchangeable structure and an original material testimony as opposed to later additions. Many examples testify to this point. Ouro Preto was considered "an unequalled heritage because of its simple but original urban architecture."[178] Røros' "remarkably complete state of preservation" was validated by the virtually unchanging views, which were visible in an engraving of the town dating from the 1860s.[179] The ideal of a city untouched by modernity becomes perhaps most apparent in the case of Valletta, which was appreciated as one of the "rare urban inhabited sites which has preserved in near entirety its original features."[180] This attitude was also reflected at the level of the terminology that was used, as some cities were described as "museum cities" or "conservatories,"[181] even though the creation of city museums was never encouraged by ICOMOS.[182] The use of vocabulary, such as "preservation," "maintenance" or "complete state," has also served to underline the original state and stability of the evaluated sites.[183]

The threat of change has acted as the prime catalyst for urban conservation at all times.[184] The ideal of unchangeable cities in the context of World Heritage was a natural response to post-war urbanism. As the World Heritage Convention was essentially defined in the context of threat to places – threat often equaling destruction and demolition – and as its later implementation saw a similar response to large-scale development projects, it is hardly surprising that priority was given to underlining the stability of cultural heritage and to depicting all change as negative. As a by-product of this concern, however, the more positive aspects of urban dynamics and continuity were largely left unidentified.[185]

This kind of equating of change with threat and demolition is well illustrated by the ICOMOS commentary in 1986 regarding the Old City of Sana'a. After explaining successive reconstructions of Sana'a under Ottoman domination and the relative recentness of the houses in the old city ("most are barely two centuries old"), ICOMOS continued, stating that "[p]reserved from change throughout the centuries and until 1962, the historic city of Sana'a, like most medinas today, increasingly suffers from facilities and an architectural style poorly adapted to living conditions in a modern capital. Many dangers threaten Sana'a."[186] While there is no reason to deny the disruptiveness of the post-1960s change in Sana'a, it is nevertheless striking that Sana'a was perceived of as something that had been preserved from all change until 1962. In this context the word 'change' constituted only negative recent change.

The same ideal of an unchangeable city was also promoted by the guidelines discussing the inclusion of historic towns in the World Heritage List, the drafting of which had been a direct response to nominations of big cities that the Committee had considered problematic, especially those concerning Rome and Jerusalem in the early 1980s. The solution was to encourage nominations and inscriptions of small or medium-sized towns "which are in a position to manage any potential growth, rather than the great metropolises, which cannot readily provide files that will serve as a satisfactory basis for their inclusion as complete units." As part of the guidelines, the outstanding universal value of World Heritage cities, or towns – the word 'city' was not used at all, most likely in order to underline the small town approach – was founded on architectural values, unchanging character and lack of changes brought about by industrialization and later urbanization. Even though the development of historic towns under the influence of socioeconomic and cultural change was not disregarded altogether, it was seen as highly problematic by nature.[187]

The urban heritage guidelines, although undoubtedly influencing many states' nomination policies, were not able to prevent the large-city nominations altogether.

Istanbul was inscribed on the World Heritage List in 1985 followed by Aleppo, Budapest, Mexico City, Timbuktu, Djenné, St. Petersburg and Paris in the late 1980s and early 1990s. It should also be noted that even in cases of perceived disruptive change, ICOMOS rarely recommended rejection of the city in question. More often, a recommendation of inscription was eventually given regardless of the perceived "irreversible change" and inadequate conservation policies in force, but usually with some conditional requirements concerning future urban planning. Often on these occasions ICOMOS encouraged a sectoral or monumental delimitation of the World Heritage area. Timbuktu and Istanbul serve as examples of this realistic, but limited approach. The two successive nominations by the Malian government in 1979 and 1987 comprised the entire Old Town of Timbuktu, but the World Heritage Committee, based on the recommendation by ICOMOS, wanted to restrict the World Heritage site in such a way as to include only the three great mosques, the 16 cemeteries and the mausolea of Timbuktu. These were considered a sufficient example of the historical 'Golden Age' in the city.[188] The historical importance of Istanbul, together with its present status as a large metropolis causing serious threats to the historical and cultural heritage of the old town, provided a context in which "the proposal for inclusion must be examined. Its restrictive nature illustrates the recent deterioration of the urban fabric, but also the political will to safeguard a number of privileged sites."[189]

Coming from the core of the Western urban protection movement, Lübeck was without doubt subject to one of the most critical evaluations ever given by ICOMOS with regard to cities. The Hanseatic city was first proposed for inclusion in the List in 1983. In its subsequent negative statement, ICOMOS denounced the "policy of cutting into the urban centre" and "the very selective reconstruction" that had been pursued in the city up until very recently; for while this policy had "permitted the replacement of the most important churches and monuments" it was "sometimes in brutal violation of the historic plan of the city and of its structure, which was that of small parceling."[190] The Bureau of the World Heritage Committee, however, was not ready to reject the nomination altogether but proposed a deferral on the basis of an insufficiently defined perimeter of protection in the nomination file, and because it considered that Lübeck needed to be "included in a global historical perspective of Hanseatic cities."[191] The Federal Republic of Germany decided to withdraw the nomination before it was to be handled in the 1983 Committee meeting.[192] The re-nomination of this site by Germany in 1987 was based on a more restricted delineation of the World Heritage area, composed of three separate zones and excluding the commercial center, which after World War II had been exposed to the most extensive redevelopment. Having taken note that the new proposal conformed to the wishes expressed in 1983 by the Bureau regarding the redefinition of the parameter of protection, ICOMOS, in its own words, could "only give a favorable opinion." The new nomination was considered "satisfactory in form if not in spirit,"[193] leading to the inclusion of the *Altstadt* in the World Heritage List in 1987.

Here we should take a moment to compare cities with monumental heritage. As argued by Sophia Labadi, the States Parties, when making World Heritage nominations, have used the concept of authenticity primarily in the meaning of the "original state and form" in contrast to "authenticity as a changing process" reflective of subsequent stages in relation to the site's historical time line. According to Labadi this holds true both for the pre- and post-Nara era.[194] As we have seen, ICOMOS, in reference to cities, also used the authenticity concept to presuppose original state, particularly when describing the authenticity of individual buildings in the city.[195] In the overall usage, however, the original state was a more flexible concept than one limited only to the time of creation. Due to the very nature of cities as changing embodiments, the ideal of unchangeable cities needed to be adjusted to fit various situations, and, as a result, a certain flexibility in the consideration of change and authenticity followed. There was an understanding of cities as bearing witness to a "complex authenticity."[196] So even though ICOMOS idealized original state and wanted to exclude the parameter of change from the definition of World Heritage cities, it in many cases had to deal with it.

It should also be pointed out that even during the early years of the implementation of the Convention the establishment of outstanding universal value of a city in relation to its historical time line and to subsequent developments could acquire very different meanings, depending on the selected perspective. This is well illustrated by the conclusions come to in two successive and otherwise identical ICOMOS evaluations concerning Bern in 1983:

> However, ICOMOS is of the opinion that the successive modifications which the old city of Bern has undergone do not militate in favour of its choice as a typical example of medieval urbanism.[197]
>
> While taking account of the significant modifications that have been made since its foundation in the 12th century, ICOMOS considers that it constitutes a positive example of how a medieval urban structure can be adapted to fulfill functions which are increasingly complex, notably the function of a capital city of a modern state.[198]

This example, together with an analogical example concerning Puebla,[199] shows how a city with a 'typical face' could be considered as representative of urban continuity thanks to a slight alteration of focus concerning the main narrative. The major nineteenth-century change could be integrated as part of the description of outstanding universal value. These changes of perspective were required by the Bureau of the World Heritage Committee; this thus suggests that it was the Committee, not ICOMOS, that allowed more flexibility with respect to continuity.

The early implementation of the World Heritage Convention saw the advancement of a static image of the city. How, then, was change treated in more recent ICOMOS evaluations, in particular after the adoption of the Vienna Memorandum? During this period the underlying ideal has also been an unchanged city. Still, as in the earlier period, we may find references to recent physical, structural or visual change, which ICOMOS has deplored but accepted as a necessary evil, especially if this change has recently been made impossible by local conservation regulations,[200] or if the disruptive change has been visually masked so as to not be able to be perceived from the World Heritage area.[201] Recently, however, the kind of change that ICOMOS has considered disruptive to the authenticity and integrity of the nominated city has perhaps more easily led to a negative statement by the organization than in the early period.[202]

There has been a general acceptance of gradual change over time and of "normal urban dynamics" as part of continuity,[203] but usually there has also been a marked separation of the two concepts of continuity and change: while continuity has been regarded as a positive force, change has been perceived almost solely as meaning uncontrolled growth and insensitive building. In other words, change, in tandem with permanency, has rarely been granted the status of a productive element for continuity.[204] What is more, stagnation, even neglect at some point of urban development, and the consequent lack of physical change since that time have also continued to function as key guarantees of outstanding universal value even within the context of the more recent evaluation texts. Several examples could be cited to make this attitude explicit, among them this authenticity statement concerning Panama: "The urban layout of the Historic District of Panama City may be considered to be entirely authentic, preserving its original form unchanged. The organically developed stock of buildings from the 18th to 20th centuries has been little changed over time, largely owing to neglect."[205] This is not to dispute the fact that stagnation in many cases has guaranteed the actual survival of the urban built fabric, or that material permanency should not be an important consideration in the context of protected cities, but is simply to make visible the constraints in the articulation of continuity.

Ambiguity in relation to change may also be discerned in the ICOMOS evaluation concerning Le Havre, which, on the one hand, recognized development and change as part of the "living city," but, on the other hand, complimented the fact that there had been as few physical changes as possible.[206] The statement by ICOMOS that Tel Aviv is not a "'town-museum,' but a city where tension between 'living city' and 'maintaining the present state' continues to exist,"[207] is illustrative of similar ambiguity. While it is a statement suggesting that dynamic and large urban settlements may also become designated as World Heritage, it implies that the "living city" and "maintaining the present state" should be treated as two separate elements, and not as part of the same process, and that there exists a necessary "tension" between the two.

These views are at odds with the notion that even though culture becomes concrete in physical structures, it is human activity that is essential to its evolvement, and that this continuity of culture necessarily involves some concessions to material permanency.[208] Accordingly, authenticity lies not only in the past or in the physical structures of the built heritage, but in the present and in the communities' (intangible) association with the site, [209] and "in the continuation of the evolution and development of society."[210] An interesting discussion within the framework of ICOMOS evaluations involved rooftop additions in certain listed buildings in Tel Aviv. These were first introduced by ICOMOS as a problem needing special attention, but, later in the text, accepted, "within certain limits," as something that represented "traditional continuity" in the Jewish context of the Diaspora.[211]

Here it warrants returning to discuss briefly the use of criterion v, which refers to traditional human settlements undergoing irreversible change. This criterion does not in itself exclude change or the idea of a continuous culture. Still, the usage of the criterion over the years suggests that the traditional human settlement has usually been treated by the Committee and ICOMOS as something anti-modern and stagnant. This resembles the perhaps unintentional inclination in the overall UNESCO framework to represent indigenous peoples as distinctively pre-modern and "frozen in time" even though constituting living cultures.[212] Especially when inseparably linked to the concern with irreversible change, as has been common in the World Heritage framework, the traditional human settlement becomes articulated as a stagnant form of life. The practical application of the criterion in the case of cities – the majority of which are small and vernacularly built – also implies that the criterion was not designed for the purposes of changing urban communities. On the other hand, and knowing that according to the World Heritage organization urban heritage – more today than ever before – is under the impact of irreversible change,[213] the decreasing importance of criterion v in the context of World Heritage cities seems surprising.

In certain cases, however, aspects of continuity and change have been integrated as an essential part of the assessment of the outstanding universal value of historic cities, such as in the case of San Cristóbal de la Laguna and Vienna:

> A living town has a dynamic which results in a continuing process of modification, and this dynamic is in itself an aspect of authenticity. This is well illustrated in San Cristóbal de la Laguna, which has evolved continuously since its foundation five hundred years ago. This can be 'read' in its street pattern, its open spaces, and its monuments, which preserve a visible continuity. This results, somewhat paradoxically, from its relative economic backwardness over the past two hundred years, which has prevented the wholesale destruction of much of its designed and built urban fabric. The town therefore has an unassailable authenticity in this respect. In terms of detail, the authenticity is high. Original facades survive in large numbers, providing an authentic historic streetscape which demonstrates the diverse origins of the town's architecture. […][214]
>
> The historic town of Vienna is an example of a town where the city centre has always remained at the same place. As a result, the city has undergone a continuous process of change, rehabilitation, and adaptation. […] Through this process of gradual change and development, the town has acquired its particular integrity and specific character and its outstanding value.[215]

Both of these examples nevertheless include a paradox. In the case of San Cristóbal de la Laguna this paradox involves, as also indicated in the evaluation text itself, determining the authenticity resulting from the dynamic nature of the city by actual stagnation over the past two hundred years, and ending up highlighting the authenticity of material and design. In the Vienna case it is paradoxical that it was this very city, subject to a continuous, and, at the time of its World Heritage nomination, a generally accepted process of gradual change and development, in which a dispute arose over a planned high-rise building. The Vienna example suggests that it is challenging to go beyond the point of traditional perspectives with regard to the outstanding universal value of cities in relation to their dynamics, development and authenticity, and that when doing so, there is a problem with opening the door to potentially out-of-place-and-scale plans in the urban context.[216] This has been the main challenge of the World Heritage work in relation to cities in the past, and, most likely, will also be the main challenge in the future.

It seems that even though ICOMOS, after Nara, has in certain cases articulated the idea that each city "has its own authenticity,"[217] it has nevertheless continuously treated the concept of authenticity as a universal quality measurable according to predefined international rules. In light of the post-Nara ICOMOS evaluations it also seems evident that the verification process related to defining authenticity has not become any easier with the – at least apparent – availability of multiple variables.[218] As the World Heritage organization continues to find authenticity assessment for World Heritage cities important, the question remains as to why it should not treat cities consistently in the language applied to cultural landscapes for which the definition of authenticity relates to "their distinctive character and components," rather than equate them with individual buildings and monumental heritage. It is difficult to say whether this would mean any significant change; surprisingly, the few existing cases of the treatment of urban nominations as sites or cultural landscapes also actually display either a rather traditional view as to authenticity or an undynamic view.[219] Finally, at present, the concept of authenticity is in a state of transition within the official discourses on World Heritage. The introduction of the concept of integrity as another qualifying condition for cultural heritage since the mid-2000s has had its influence on the use of the authenticity concept. With regard to several recent urban nominations we find integrity and authenticity considered in such a way as to make it difficult to separate these two concepts from one another, both revolving around what used to be called material authenticity. For instance, the "integrity index," introduced by Switzerland and praised by ICOMOS in association with the nomination of La Chaux-de-Fonds and Le Locle, was used to show the percentage of pre-1930 buildings in these two cities, which had preserved unaltered.[220] As a result, the range of reflection on authenticity may again be declining. As with many other concepts within the official World Heritage discourse, there is a genuine attempt to depart from the most traditional understandings of authenticity while still clinging to the very same interpretations.

Notes

1   ICOMOS evaluation for the nomination of the World Heritage property, "Rome," Italy, No 91, 1980.
2   ICOMOS evaluation for the nomination of the World Heritage property, "Harar Jugol," Ethiopia, No 1189 rev, 10 April 2006.
3   G. J. Ashworth and J. E. Tunbridge, *The Tourist-historic City: Retrospect and Prospect of Managing the Heritage City* (Amsterdam: Pergamon, 2000); Peter J. Larkham,

Conservation and the City (London: Routledge, 1996); UNESCO, *Recommendation concerning Safeguarding and Contemporary Role of Historic Areas*, 1976.

4 Labadi, *UNESCO*, 61.
5 Smith, *Uses of Heritage*. See also Hobsbawn, "Introduction: inventing traditions;" Korhar, *Germany's Transient Pasts*.
6 ICOMOS evaluation for the nomination of the World Heritage property, "Ouro Preto," Brazil, No 124, May 1980. See also ICOMOS evaluation for the nomination of the World Heritage property, "Toledo," Spain, No 379, April 1986.
7 ICOMOS evaluation for the nomination of the World Heritage property, "Zamosc," Poland, No 564, October 1992. See also ICOMOS evaluation for the nomination of the World Heritage property, "Lübeck," Germany, No 272, May 1983.
8 ICOMOS evaluation for the nomination of the World Heritage property, "Valletta," Malta, No 131, May 1980. See also ICOMOS evaluation for the nomination of the World Heritage property, "Avila," Spain, No 348 rev., November 1985; ICOMOS evaluation for the nomination of the World Heritage property, "Siena," Italy, No 717, September 1995.
9 ICOMOS evaluation for the nomination of the World Heritage property, "Brasilia," Brazil No 445, October 1987.
10 See, for example, ICOMOS evaluation for the nomination of the World Heritage property, "Budapest," Hungary, No 400, April 1987; ICOMOS evaluation for the nomination of the World Heritage property, "Istanbul," Turkey, No 356, July 1985.
11 ICOMOS evaluation for the nomination of the World Heritage property, "Olinda," Brazil, No 189, May 1982.
12 ICOMOS evaluation for the nomination of the World Heritage property, "Plovdiv," Bulgaria, No 218, June 1983. World Heritage Centre, World Heritage Sites, Nomination files (1978–1999), Rejected (CD in author's possession).
13 UNESCO, World Heritage Committee, Seventh Session (Florence, Italy, 5–9 December 1983), Report of Rapporteur, C/83/CONF.009/8, Paris, January 1984), p. 11, http://whc.unesco.org/archive/1983/sc-83-conf009-8e.pdf.
14 ICOMOS evaluation for the nomination of the World Heritage property, "Old City and Ramparts of Alanya," Turkey, No 1354, 10 March 2011. UNESCO, World Heritage Committee, 35th Session (UNESCO, June 2011), ICOMOS Evaluations of Nominations of Cultural and Mixed Properties, p. 307–316, http://whc.unesco.org/archive/2011/whc11-35com-inf.8B1e.pdf.
15 Ron van Oers, "Preventing the Goose with the Golden Eggs from catching Bird Flu – UNESCO's efforts in Safeguarding the Historic Urban Landscape." Keynote paper for the 42nd Congress of the International Society of City and Regional Planners (ISoCaRP). *Cities between Integration and Disintegration*, 14–18 September, 2006, Istanbul, Turkey.
16 See also Cleere, "The Uneasy Bedfellows," 25.
17 See, for example, ICOMOS evaluation for the nomination of the World Heritage property, "Mantua and Sabbioneta," Italy, No 1287, 11 March 2008.
18 For a discussion concerning the use of criterion i and fine arts understanding of cultural heritage see also Bandarin and Labadi, *World Heritage,* 95.
19 ICOMOS evaluation for the nomination of the World Heritage property, "Kairouan," Tunisia, No 499, May 1988. For similar statements see also ICOMOS evaluation for the nomination of the World Heritage property, "Marrakesh," Morocco, No 331, July 1985; ICOMOS evaluation for the nomination of the World Heritage property, "Sana'a," Yemen, No 385, April 1986; ICOMOS evaluation for the nomination of the World Heritage property, "Ghadamès," Libyan Arab Jamahiriya, No 362, April 1986; ICOMOS evaluation for the nomination of the World Heritage property, "Sousse," Tunisia, No 498, July 1988.
20 King, "Terminologies," 141.

21 ICOMOS evaluation for the nomination of the World Heritage property, "Cuenca," Ecuador No 863, September 1999.
22 For Old Rauma see Chapter 6.
23 ICOMOS evaluation for the nomination of the World Heritage property, "Vigan," Philippines, No 502, September 1989. World Heritage Centre, World Heritage Sites, Nomination files (1978–1999), Rejected (CD in author's possession).
24 ICOMOS evaluation for the nomination of the World Heritage property, "Vigan," Philippines, No 502rev, September 1999.
25 Feilden and Jokilehto, *Management Guidelines*, 78.
26 ICOMOS evaluation for the nomination of the World Heritage property, "Arequipa," Peru, No 1016, September 2000.
27 ICOMOS evaluation for the nomination of the World Heritage property, "Lyons," France, No 872, October 1998; ICOMOS evaluation for the nomination of the World Heritage property, "Noto," Italy, No 1024rev, January 2002; ICOMOS evaluation for the nomination of the World Heritage property, "Episcopal City of Albi," France, No 1337, 17 March 2010.
28 Operational Guidelines, 8 July 2015, Annex 3.
29 ICOMOS evaluation for the nomination of the World Heritage property, "Salvador de Bahia," Brazil, No 309, July 1985.
30 Dicks, *Culture*, 17, 43; Urry, *The Tourist Gaze*, 145–149.
31 For discussion see, for example, D. Jacques and P. Fowler, "Conservation of landscapes in post-industrial countries," in *Cultural Landscapes of Universal Value – Components of a Global Strategy*, ed. B. von Droste, H. Plachter and M. Rössler (Jena: Gustav Fischer Verlag, 1995), 414; Larkham, *Conservation and the City*, 6, 22; David Crouch, "The perpetual performance and emergence of heritage," in *Culture, Heritage and Representation: Perspectives on Visuality and the Past*, ed. Emma Waterton and Steve Watson (Surrey: Ashgate, 2010), 57–71.
32 Jokilehto, Jukka, "Considerations on authenticity and integrity in world heritage context." *City & Time* 2: 1 (2006): 3, accessed April 24, 2012, www.ct.ceci-br.org.
33 Steve Watson and Emma Waterton, "Introduction: a visual heritage," in *Culture, Heritage and Representation: Perspectives on Visuality and the Past*, ed. Emma Waterton and Steve Watson (Surrey: Ashgate, 2010), 2.
34 ICOMOS evaluation for the nomination of the World Heritage property, "Segovia," Spain, No 311rev., November 1985; ICOMOS evaluation for the nomination of the World Heritage property, "San Gimignano," Italy, No 550, April 1990; ICOMOS evaluation for the nomination of the World Heritage property, "Kutná Hora," Czech Republic, 732, September 1995; ICOMOS evaluation for the nomination of the World Heritage property, "Salzburg," Austria, No 784, October 1996.
35 ICOMOS evaluation for the nomination of the World Heritage property, "Cairo," Egypt No 89, 1979.
36 ICOMOS evaluation for the nomination of the World Heritage property, "Old City of Sana'a," Yemen, No 385, April 1986; ICOMOS evaluation for the nomination of the World Heritage property, "Edinburgh," United Kingdom, No 728, September 1995.
37 ICOMOS evaluation for the nomination of the World Heritage property, "Sibiu, the Historic Centre," Romania, No 1238, 21 January 2007. UNESCO, World Heritage Committee, Thirty-first ordinary session, Christchurch, New Zealand, 23 June – 2 July, 2007, Evaluations of cultural properties, WHC-07/31.COM/INF.8B1, p. 199–204, http://whc.unesco.org/archive/2007/whc07-31com-inf8b1e.pdf.
38 ICOMOS evaluation for the nomination of the World Heritage property, "Goiás," Brazil, No 993, September 2001. My emphasis.
39 For a similar argument concerning the Vienna Memorandum see Smith, "Marrying," 67.
40 On objectification of environment see Arnold Berleant, *Art and Engagement* (Philadelphia: Temple University Press, 1991), 82–88; Elisa El Harouny,

"Historiallinen puukaupunki – tulevaisuuden elävää kaupunkiympäristöä?" in *Eletty ja muistettu tila*, ed. Taina Syrjämaa and Janne Tunturi (Helsinki: Finnish Literature Society, 2002), 261–262; Crouch, "The perpetual," 57.

41 Edward Relph, *Place and Placelessness* (London: Pion, 1976); Tim Cresswell, *Place. A Short Introduction* (Malden: Blackwell, 2004).

42 Smith, "Marrying," 67.

43 Ibid., 68.

44 ICOMOS evaluation for the nomination of the World Heritage property, "Quito," Ecuador, No 2, Paris, 7 June 1979, Ernst Connally, Secretary General to Mr. Firouz Bagerzadeh, Chairman, World Heritage Committee. World Heritage Centre, World Heritage Sites, Nomination files (1978–1997), The Americas (CD in author's possession). See also ICOMOS evaluation for the nomination of the World Heritage property, "Sana'a," Yemen, No 385, April 1986; ICOMOS evaluation for the nomination of the World Heritage property, "Cienfuegos," Cuba, No 1202, April 2005.

45 See, for example, ICOMOS evaluation for the nomination of the World Heritage property, "Luxembourg," Luxembourg, No 699, October 1994. For exceptions to this rule, see, ICOMOS evaluation for the nomination of the World Heritage property, "Potosi," Bolivia, No 420, April 1987; ICOMOS evaluation for the nomination of the World Heritage property, "Guanajuato," Mexico, No 482, September 1988; ICOMOS evaluation for the nomination of the World Heritage property, "Trinidad and the Valley de los Ingenios," Cuba, No 460, September 1988.

46 ICOMOS evaluation for the nomination of the World Heritage property, "Lijiang," China, No 811, September 1997.

47 ICOMOS evaluation for the nomination of the World Heritage property, "Valletta," Malta, No 131, May 1980. See also ICOMOS evaluation for the nomination of the World Heritage property, "San Miguel and the Sanctuary of Atotonilco," Mexico No 1274, 11 March 2008.

48 ICOMOS evaluation for the nomination of the World Heritage property, "Dresden Elbe Valley," Germany, No 1156, March 2004. See also ICOMOS evaluation for the nomination of the World Heritage property, "Lijiang," China, No 811, September 1997; ICOMOS evaluation for the nomination of the World Heritage property, "Karlskrona," Sweden, No 871, October 1998; ICOMOS evaluation for the nomination of the World Heritage property, "Riga," Latvia, No 852, September 1997.

49 ICOMOS evaluation for the nomination of the World Heritage property, "Zamosc," Poland, No 564, October 1992.

50 Waterton, Smith and Campbell, "The Utility of Discourse Analysis," 344.

51 Larkham, *Conservation and the City*, 24–25.

52 On postmodernist historicism see Gavriel D. Rosenfeld, *Munich and Memory: Architecture, Monuments, and the Legacy of the Third Reich* (Berkley: University of California Press, 2000), 239–254; Wim Denslagen *Romantic Modernism: Nostalgia in the World of Conservation* (Amsterdam: Amsterdam University Press, 2009).

53 For Old Rauma see Chapter 6.

54 ICOMOS evaluation for the nomination of the World Heritage property, "Ouro Preto," Brazil, No 124, May 1980; ICOMOS evaluation for the nomination of the World Heritage property, "Bordeaux," France, No 1256, 21 January 2007. See also, for example, ICOMOS evaluation for the nomination of the World Heritage property, "Évora," Portugal, No 361, April 1986; ICOMOS evaluation for the nomination of the World Heritage property, "Val di Noto," Italy, No 1024rev, January 2002; ICOMOS evaluation for the nomination of the World Heritage property, "The Canal Area of Amsterdam," Netherlands, No 1349, 17 March 2010.

55 ICOMOS evaluation for the nomination of the World Heritage property, "Valparaíso," Chile No 959rev, March 2003.

56 ICOMOS evaluation for the nomination of the World Heritage property, "Bordeaux," France, No 1256, 21 January 2007.

57 ICOMOS evaluation for the nomination of the World Heritage property, "Vilnius," Lithuania, No 541, October 1994.

58 ICOMOS evaluation for the nomination of the World Heritage property, "Melaka and George Town," Malaysia, No 1223, 11 March 2008.

59 van Oers, "Preventing the Goose."

60 *World Heritage Convention* 1972, Article 1.

61 van Oers, "Managing cities," 13.

62 UNESCO, World Heritage Committee, Thirty-fourth Session, Brasilia, Brazil, 25 July – 3 August 2010, Item 7.1 of the Provisional Agenda: Historic Urban Landscape, WHC-10/34.COM/7.1 (Paris, 18 June 2010), p. 2, http://whc.unesco.org/archive/2010/whc10-34com-7.1e.pdf.

63 UNESCO, World Heritage Committee, Thirty-fourth Session, Brasilia, Brazil, 25 July – 3 August 2010, Item 13 of the Provisional Agenda: Revision of the Operational Guidelines, WHC-10/34.COM/13 (Paris, 18 June 2010), p. 26, http://whc.unesco.org/archive/2010/whc10-34com-13e.pdf.

64 Operational Guidelines, 8 July 2015, Annex 3.

65 Ibid., Article 88.

66 See for, example, ICOMOS evaluation for the nomination of the World Heritage property, "Provins," France, No 873rev, September 2001; ICOMOS evaluation for the nomination of the World Heritage property, "Stralsund and Wismar," Germany, No 1067, January 2002 ICOMOS evaluation for the nomination of the World Heritage property, "Úbeda-Baeza," Spain, No 522rev bis, March 2003; ICOMOS evaluation for the nomination of the World Heritage property, "Třebíč," Czech Republic, No 1078, March 2003; ICOMOS evaluation for the nomination of the World Heritage property, "Episcopal City of Albi," France, No 1337, 17 March 2010.

67 ICOMOS evaluation for the nomination of the World Heritage property, "San Marino," San Marino, No 1245, 11 March 2008. See also ICOMOS evaluation for the nomination of the World Heritage property, "Třebíč," Czech Republic, No 1078, March 2003; ICOMOS evaluation for the nomination of the World Heritage property, "Tel Aviv," Israel, No 1096, March 2003; ICOMOS evaluation for the nomination of the World Heritage property, "The Canal Area of Amsterdam," Netherlands, No 1349, 17 March 2010.

68 Operational Guidelines, 8 July 2015, Article 89.

69 For different degrees of integrity see Jokilehto, "Considerations," 14; Herb Stovel, "ICOMOS Position Paper," in *World Heritage and Buffer Zones. International Expert Meeting on World Heritage and Buffer Zones, Davos, Switzerland 11 – 14 March 2008*, ed. Oliver Martin and Giovanna Piatti (Paris: UNESCO World Heritage Centre, 2009), 27.

70 ICOMOS evaluation for the nomination of the World Heritage property, "Bridgetown and it Garrison," Barbados, No 1376, 10 March 2011. See also ICOMOS evaluation for the nomination of the World Heritage property, "Historical City of Jeddah," Saudi Arabia, No 1361, 10 March 2011. UNESCO, World Heritage Committee, 35th Session (UNESCO, June 2011), ICOMOS Evaluations of Nominations of Cultural and Mixed Properties, p. 85–95, http://whc.unesco.org/archive/2011/whc11-35com-inf.8B1e.pdf.

71 UNESCO, World Heritage Committee, Eighteenth Session (Phuket, Thailand, 12–17 December 1994), Report of the Expert Meeting on the "Global Strategy" and Thematic Studies for a Representative World Heritage List (UNESCO Headquarters, 20–22 June 1994), WHC.94/CONF.003/INF.06, Paris, 13 October 1994, http://whc.unesco.org/archive/1994/whc-94-conf003-inf6e.pdf.

72 van Oers, "Preventing the Goose."

73 Fowler, "World Heritage Cultural Landscapes," 22.

74 ICOMOS evaluation for the nomination of the World Heritage property, "Buenos Aires," Argentina, No 1296, 11 March 2008. UNESCO, World Heritage Committee,

Thirty-Second Ordinary Session, Quebec City, Canada, 2–10 July, 2008, ICOMOS Evaluations, WHC.08/32.COM/INF.8B1, p. 236–247, http://whc.unesco.org/archive/2008/whc08-32com-inf8B1e.pdf.

75 The first Operational Guidelines in 1977 defined buffer zones as "natural or man-made surroundings that influence the physical state of the property or the way in which the property is perceived." Operational Guidelines, 20 October 1977, Article 25. For the current wording see Operational Guidelines, 8 July 2015, Articles 103–107.

76 See also Stovel, "ICOMOS Position Paper," 23–24.

77 ICOMOS evaluation for the nomination of the World Heritage property, "Venice and its Lagoon," Italy, No 394, May 1987.

78 ICOMOS evaluation for the nomination of the World Heritage property, "Leningrad," Soviet Union, No 540, April 1990.

79 See, for example, ICOMOS evaluation for the nomination of the World Heritage property, "San Gimignano," Italy, No 550, April 1990.

80 See, for example, ICOMOS evaluation for the nomination of the World Heritage property, "Salvador de Bahia," Brazil, No 309, July 1985; ICOMOS evaluation for the nomination of the World Heritage property, "Paris, Banks of the Seine," France, No 600, May 1991.

81 ICOMOS evaluation for the nomination of the World Heritage property, "Assisi," Italy, No 990, September 2000.

82 ICOMOS evaluation for the nomination of the World Heritage property, "Bordeaux," France, No 1256, 21 January 2007.

83 ICOMOS evaluation for the nomination of the World Heritage property, "Melaka and George Town," Malaysia, No 1223, 11 March 2008.

84 ICOMOS evaluation for the nomination of the World Heritage property, "Harar Jugol," Ethiopia, No 1189 rev, 10 April 2006.

85 ICOMOS evaluation for the nomination of the World Heritage property, "San Marino," San Marino, No 1245, 11 March 2008. See also ICOMOS evaluation for the nomination of the World Heritage property, "La Chaux-de-Fonds/Le Locle," Switzerland, No 1302, 10 March 2009.

86 ICOMOS evaluation for the nomination of the World Heritage property, "Úbeda-Baeza," Spain, No 522rev bis, March 2003.

87 ICOMOS evaluation for the nomination of the World Heritage property, "La Plata," Argentina, No 979, 21 January 2007. UNESCO, World Heritage Committee, Thirty-first Session, Christchurch, New Zealand, 23 June – 2 July, 2007, Evaluations of cultural properties, WHC-07/31.COM/INF.8B1, p. 251–258, http://whc.unesco.org/archive/2007/whc07-31com-inf8b1e.pdf.

88 *Venice Charter* 1964, Article 6.

89 Concerning borders reflecting power relations see Anssi Paasi "Rajat ja identiteetti globalisoituvassa maailmassa," in *Eletty ja muistettu tila*, ed. Janne Tunturi and Taina Syrjämaa (Helsinki: Finnish Literature Society, 2002), 159.

90 C. Berkowitz and J. Hoffmann, "The White City of Tel-Aviv," in *World Heritage and Buffer Zones*. World Heritage Papers 25, ed. Oliver Martin and Giovanna Piatti (Paris: World Heritage Centre, 2007), 125–128.

91 Stovel, "ICOMOS Position Paper," 28; *Xi'an Declaration on the Conservation of the Setting of Heritage Structures, Sites and Areas*. Xi'an, China: 15th General Assembly of ICOMOS, 2005, accessed February 15, 2016, www.icomos.org/charters/xian-declaration.pdf; *Partnerships for World Heritage Cities: Culture as a Vector for Sustainable Urban Development*. World Heritage Papers 9 (Paris: World Heritage Centre, 2003); Ron van Oers and Sachiko Haraguchi, eds., *Managing Historic Cities*. World Heritage Papers 27 Paris: World Heritage Centre/UNESCO, 2010.

92 Jokilehto, "Considerations," 6.

93 ICOMOS evaluation for the nomination of the World Heritage property, "Marrakesh," Morocco, No 331, July 1985. My emphasis. See also ICOMOS evaluation for the nomination of the World Heritage property, "M'Zab Valley," Algeria, No 183, May 1982.

94 Nomination to World Heritage List, Djenné, Mali, February 1979. World Heritage Centre, World Heritage Sites, Nomination files (1978–1997), Africa and the Arab States (CD in author's possession).

95 ICOMOS evaluation for the nomination of the World Heritage property, "Djenné," Mali, No 116, May 1981.

96 ICOMOS evaluation for the nomination of the World Heritage property, "Djenné," Mali, No 116 Rev., May 1988. As part of establishing architectural value, ICOMOS nevertheless described the less-monumental and less-elitist domestic architecture in Djenné.

97 See also Nomination to World Heritage List, Shibam, Yemen, November 11, 1981. World Heritage Centre, World Heritage Sites, Nomination files (1978–1997), Africa and the Arab States (CD in author's possession); ICOMOS evaluation for the nomination of the World Heritage property, "Shibam," Yemen, No 192, May 1982. For Old Rauma see Chapter 6.

98 ICOMOS evaluation for the nomination of the World Heritage property, "Goiás," Brazil, No 993, September 2001.

99 ICOMOS evaluation for the nomination of the World Heritage property, "Vienna," Austria, No 1033, September 2001.

100 See, however, ICOMOS evaluation for the nomination of the World Heritage property, "Old City and Ramparts of Alanya," Turkey, No 1354, 10 March 2011. UNESCO, World Heritage Committee, 35th Session (UNESCO, June 2011), ICOMOS Evaluations of Nominations of Cultural and Mixed Properties, p. 307–316, http://whc.unesco.org/archive/2011/whc11-35com-inf.8B1e.pdf.

101 See also ICOMOS evaluation for the nomination of the World Heritage property, "Mostar," Bosnia and Herzegovina, No 946 rev, April 2005. For other kinds of connections see ICOMOS evaluation for the nomination of the World Heritage property, "San Miguel and the Sanctuary of Atotonilco," Mexico, No 1274, 11 March 2008, which mentions that some of the eighteenth-century "residences are still inhabited by descendants of families that constructed them."

102 ICOMOS evaluation for the nomination of the World Heritage property, "Macao," China, No 1110, April 2005. See also ICOMOS evaluation for the nomination of the World Heritage property, "Melaka and George Town," Malaysia, No 1223, 11 March 2008.

103 Gibson, "Cultural Landscapes and Identity," 73.

104 ICOMOS evaluation for the nomination of the World Heritage property, "Le Havre," France, No 1181, April 2005.

105 ICOMOS evaluation for the nomination of the World Heritage property, "Bardejov," Slovakia, No 973, September 2000.

106 ICOMOS evaluation for the nomination of the World Heritage property, "Melaka and George Town," Malaysia, No 1223, 11 March 2008.

107 For the non-European heritage conservation approach and the tensions within UNESCO see Logan, "Globalizing Heritage," 55.

108 For an exception, see ICOMOS evaluation for the nomination of the World Heritage property, "Patmos," Greece, No 942, September 1999.

109 ICOMOS evaluation for the nomination of the World Heritage property, "Saint-Louis," Senegal, No 956, October 2000.

110 ICOMOS evaluation for the nomination of the World Heritage property, "Mantua and Sabbioneta," Italy, No 1287, 11 March 2008; ICOMOS evaluation for the nomination of the World Heritage property, "Corfu," Greece, No 978, 11 March 2007.

111 See, for example, ICOMOS evaluation for the nomination of the World Heritage property, "Melaka and George Town," Malaysia, No 1223, 11 March 2008; ICOMOS evaluation for the nomination of the World Heritage property, "Le Havre," France, No 1181, April 2005.

112 See, for example, ICOMOS evaluation for the nomination of the World Heritage property, "Goiás," Brazil, No 993, September 2001.

113 ICOMOS evaluation for the nomination of the World Heritage property, "Ibiza," Spain, No 417 rev, September 1999.

114 ICOMOS evaluation for the nomination of the World Heritage property, "Mantua and Sabbioneta," Italy, No 1287, 11 March 2008.

115 See, for example, ICOMOS evaluation for the nomination of the World Heritage property, "Willemstad," Netherlands, No 819, September 1997; ICOMOS evaluation for the nomination of the World Heritage property, "Episcopal City of Albi," France, No 1337, 17 March 2010.

116 See, for example, ICOMOS evaluation for the nomination of the World Heritage property, "Paramaribo," Surinam, No 940 rev, April 2002; ICOMOS evaluation for the nomination of the World Heritage property, "Mantua and Sabbioneta," Italy, No 1287, 11 March 2008; ICOMOS evaluation for the nomination of the World Heritage property, "Housing Estates in Berlin," Germany, No 1230, 11 March 2008.

117 Waterton, Smith and Campbell, "The Utility of Discourse Analysis," 348–350.

118 ICOMOS evaluation for the nomination of the World Heritage property, "Bridgetown and it Garrison," Barbados, No 1376, 10 March 2011.

119 ICOMOS evaluation for the nomination of the World Heritage property, "Historical City of Jeddah," Saudi Arabia No 1361, 10 March 2011. UNESCO, World Heritage Committee, 35th Session (UNESCO, June 2011), ICOMOS Evaluations of Nominations of Cultural and Mixed Properties, p. 85–95, http://whc.unesco.org/archive/2011/whc11-35com-inf.8B1e.pdf.

120 For national nomination dossiers, see Labadi, *UNESCO*, 88–89.

121 ICOMOS evaluation for the nomination of the World Heritage property, "Le Havre," France, No 1181, April 2005.

122 ICOMOS evaluation for the nomination of the World Heritage property, "Acre," Israel, No 1042, November 2001, revised 2002. My emphasis.

123 Ibid.

124 For the 1970s scholarly understanding of community see Emma Waterton and Laurajane Smith, "The recognition and misrecognition of community heritage, *International Journal of Heritage Studies*, 16: 1–2 (2010): 6–7, for quotation see page 7. For various definitions of community see Graham Crow and Graham Allan, *Community Life. An Introduction to local Social Relations* (New York: Harvester Wheatsheaf, 1994), 1–12.

125 ICOMOS evaluation for the nomination of the World Heritage property, "Harar Jugol," Ethiopia, No 1189 rev, 10 April 2006.

126 Waterton and Smith, "The recognition."

127 ICOMOS evaluation for the nomination of the World Heritage property, "Le Havre," France, No 1181, April 2005.

128 ICOMOS evaluation for the nomination of the World Heritage property, "Bridgetown and it Garrison," Barbados, No 1376, 10 March 2011.

129 ICOMOS evaluation for the nomination of the World Heritage property, "Yaroslavl," Russian Federation, No 1170, April 2005.

130 Waterton and Smith, "The recognition," 8.

131 ICOMOS evaluation for the nomination of the World Heritage property, "Cidade Velha," Cape Verde, No 1310, 10 March 2009.

132 ICOMOS evaluation for the nomination of the World Heritage property, "San Marino," San Marino, No 1245, 11 March 2008.

133 ICOMOS evaluation for the nomination of the World Heritage property, "Gdańsk," Poland, No 882, October 1998. World Heritage Centre, World Heritage Sites, Nomination files (1978–1999), Rejected (CD in author's possession).

134 ICOMOS evaluation for the nomination of the World Heritage property, "Gdańsk," Poland, No 1240, 21 January 2007. World Heritage Committee, 31st ordinary session (Christchurch, New Zealand, 23 June – 2 July), Evaluations of cultural properties, WHC-07/31.COM/INF.8B1 (2007), p. 191–198, http://whc.unesco.org/archive/2007/whc07-31com-inf8b1e.pdf.

135 ICOMOS evaluation for the nomination of the World Heritage property, "Buenos Aires," Argentina, No 1296, 11 March 2008. Evaluations of cultural properties (2008), WHC.08/32.COM/INF.8B1, p. 236–247. WHCIA.

136 ICOMOS evaluation for the nomination of the World Heritage property, "Bridgetown and its Garrison," Barbados, No 1376, 10 March 2011.

137 UNESCO, World Heritage Committee, Thirty-Fifth Session, Paris, UNESCO Headquarters (19–29 June 2011), Summary Record, WHC-11/35.COM.INF.20, 205–206, http://whc.unesco.org/archive/2011/whc11-35com-inf20.pdf.

138 ICOMOS evaluation for the nomination of the World Heritage property, "Old City and Ramparts of Alanya," Turkey, No 1354, 10 March 2011. UNESCO, World Heritage Committee, 35th Session (UNESCO, June 2011), ICOMOS Evaluations of Nominations of Cultural and Mixed Properties, p. 307–316, http://whc.unesco.org/archive/2011/whc11-35com-inf.8B1e.pdf. See also ICOMOS evaluation for the nomination of the World Heritage property, "Historical City of Jeddah," Saudi Arabia, No 1361, 10 March 2011. UNESCO, World Heritage Committee, 35th Session (UNESCO, June 2011), ICOMOS Evaluations of Nominations of Cultural and Mixed Properties, p. 85–95, http://whc.unesco.org/archive/2011/whc11-35com-inf.8B1e.pdf.

139 Gustavo Araoz, "Protecting Heritage Places Under the New Heritage Paradigm & Defining its Tolerance for Change: a Leadership Challenge for ICOMOS." Message presented to ICOMOS Executive and Advisory committees in La Valletta, Malta, October 2009, accessed April 24, 2012, www.fondazione-delbianco.org/seminari/progetti_prof/progview_PL.asp?start=1&idprog=283.

140 Michael Petzet, "Conservation or managing change." Paper presented in *Conservation Turn – Return to Conservation: Challenges and Chances in a Changing World* (Prague, 5–9 May 2010), accessed April 24, 2012, www.icomos.de/.

141 Michael Petzet, "'In the full richness of their authenticity' – The test of authenticity and the New Cult of Monuments," in *Nara Conference on Authenticity: Proceedings*, ed. Knut Einar Larsen (Paris: World Heritage Centre, 1995), 97.

142 Ito, "'Authenticity,'" 36; Jokilehto, "Authenticity," 18; MacCannell, *The Tourist*.

143 Christina Cameron, "Authenticity and the World Heritage Convention," in *Nara Conference on Authenticity: Proceedings*, ed. Knut Einar Larsen (Paris: World Heritage Centre, 1995), 283.

144 von Droste and Bertilsson, "Authenticity and World Heritage," 4–5; Henry Cleere, "The evaluation of authenticity in the context of the World Heritage Convention," in *Nara Conference on Authenticity: Proceedings*, ed. Knut Einar Larsen (Paris: World Heritage Centre, 1995), 60; Labadi, "Representations"; Labadi, *UNESCO*, 113–126.

145 Cleere, "The evaluation of authenticity," 64.

146 Operational Guidelines, January 1987, Paragraph 26.

147 See, for example, ICOMOS evaluation for the nomination of the World Heritage property, "Timbuktu," Mali, No 119 Rev, May 1988.

148 ICOMOS evaluation for the nomination of the World Heritage property, "Ghadames," Libyan Arab Jamahiriya, No 362, April 1986.

149 ICOMOS evaluation for the nomination of the World Heritage property, "Bryggen," Norway, No 59, November 1978; ICOMOS evaluation for the nomination of the World Heritage property, "Rauma," Finland, No 582, May 1991. See also ICOMOS

evaluation for the nomination of the World Heritage property, "Lijiang," China, No 811, September 1997.

150 Pressouyre, *The World Heritage Convention*, 14.

151 ICOMOS evaluation for the nomination of the World Heritage property, "Coro," Venezuela, No 658, October 1993.

152 See, for example, ICOMOS evaluation for the nomination of the World Heritage property, "Quedlinburg," Germany, No 535 Rev, October 1994; ICOMOS evaluation for the nomination of the World Heritage property, "Torun," Poland, No 835, September 1997; ICOMOS evaluation for the nomination of the World Heritage property, "Bardejov," Slovakia, No 973, September 2000; ICOMOS evaluation for the nomination of the World Heritage property, "Episcopal City of Albi," France, No 1337, 17 March 2010.

153 ICOMOS evaluation for the nomination of the World Heritage property, "Le Havre," France, No 1181, April 2005. See also ICOMOS evaluation for the nomination of the World Heritage property, "Housing Estates in Berlin," Germany, No 1230, 11 March 2008.

154 ICOMOS evaluation for the nomination of the World Heritage property, "San Marino," San Marino, No 1245, 11 March 2008; ICOMOS evaluation for the nomination of the World Heritage property, "San Miguel and the Sanctuary of Atotonilco," Mexico, No 1274, 11 March 2008.

155 Waterton, Smith and Campbell, "The Utility of Discourse Analysis," 345.

156 ICOMOS evaluation for the nomination of the World Heritage property, "Oporto," Portugal, No 755, October 1996. See also ICOMOS evaluation for the nomination of the World Heritage property, "Cordoba," Spain, No 313 bis, October 1994.

157 ICOMOS evaluation for the nomination of the World Heritage property, "Alcalá de Henares," Spain, No 876, October 1998.

158 ICOMOS evaluation for the nomination of the World Heritage property, "Verona," Italy, No 797 rev, September 2000. Jim Sajo, "Sun sets on U.S. military's 60-year stay in Verona, Italy," *Stars and Stripes*, June 13, 2004, accessed June 22, 2011, www.stripes.com/military-life/sun-sets-on-u-s-military-s-60-year-stay-in-verona-italy-1.31754.

159 ICOMOS evaluation for the nomination of the World Heritage property, "The Canal Area of Amsterdam," Netherlands, No 1349, 17 March 2010.

160 ICOMOS evaluation for the nomination of the World Heritage property, "Campeche," Mexico, No 895, September 1999.

161 ICOMOS evaluation for the nomination of the World Heritage property, "Valparaíso," Chile No 959 rev, March 2003.

162 ICOMOS evaluation for the nomination of the World Heritage property, "Melaka and George Town," Malaysia, No 1223, 11 March 2008.

163 Malaysia, Nomination Dossier, Historic Cities of the Straits of Malacca, January 2007, p. 128, http://whc.unesco.org/uploads/nominations/1223bis.pdf.

164 ICOMOS evaluation for the nomination of the World Heritage property, "Campeche," Mexico, No 895, September 1999. For a similar account see also ICOMOS evaluation for the nomination of the World Heritage property, "Camagüey," Cuba, No 1270, 11 March 2008.

165 ICOMOS evaluation for the nomination of the World Heritage property, "Popayán," Colombie, No 504, September 1989. World Heritage Centre, World Heritage Sites, Nomination files (1978–1999), Rejected (CD in author's possession).

166 Gunzburger Makaš, *Representing Competing Identities*, 106.

167 ICOMOS evaluation for the nomination of the World Heritage property, "Rhodes," Greece, No 493, September 1988; Vahtikari, "Historic cities," 112.

168 ICOMOS evaluation for the nomination of the World Heritage property, "Warsaw," Poland, No 30, June 6, 1978; Denslagen, *Romantic Modernism*, 203–204.

169 Cf. Gunzburger Makaš, *Representing Competing Identities*, 105–106.
170 ICOMOS evaluation for the nomination of the World Heritage property, "Carcassonne," France, No 345 rev, September 1997.
171 ICOMOS evaluation for the nomination of the World Heritage property, "Siena," Italy, No 717, September 1995.
172 ICOMOS evaluation for the nomination of the World Heritage property, "Pecs," Hungary, No 853, October 1998. World Heritage Centre, World Heritage Sites, Nomination files (1978–1999), Rejected (CD in author's possession).
173 ICOMOS evaluation for the nomination of the World Heritage property, "Verona," Italy, No 797 rev, September 2000.
174 ICOMOS evaluation for the nomination of the World Heritage property, "San Marino," San Marino, No 1245, 11 March 2008.
175 ICOMOS evaluation for the nomination of the World Heritage property, "Gdańsk," Poland, No 882, October 1998. World Heritage Centre, World Heritage Sites, Nomination files (1978–1999), Rejected (CD in author's possession).
176 ICOMOS evaluation for the nomination of the World Heritage property, "Mostar," Bosnia and Herzegovina, No 946 rev, April 2005.
177 ICOMOS evaluation for the nomination of the World Heritage property, "Kotor," Federative Socialist Republic of Yugoslavia No 125, October 1979.
178 ICOMOS evaluation for the nomination of the World Heritage property, "Ouro Preto," Brazil, No 124, May 1980.
179 ICOMOS evaluation for the nomination of the World Heritage property, "Røros," Norway, No 55, 1980.05. See also ICOMOS evaluation for the nomination of the World Heritage property, "Cuzco," Peru, No 273, June 1983.
180 ICOMOS evaluation for the nomination of the World Heritage property, "Valletta," Malta, No 131, May 1980.
181 ICOMOS evaluation for the nomination of the World Heritage property, "Nessebar," Bulgaria, No 217, June 1983; ICOMOS evaluation for the nomination of the World Heritage property, "Évora," Portugal, No 361, April 1986; ICOMOS evaluation for the nomination of the World Heritage property, "Toledo," Spain, No 379, April 1986; ICOMOS evaluation for the nomination of the World Heritage property, "Rauma," Finland, No 582, May 1991.
182 ICOMOS evaluation for the nomination of the World Heritage property, "Itchan Kala," Soviet Union / Uzbekistan, No 543, April 1990.
183 Labadi, *UNESCO*, 115.
184 Madgin, *Heritage, Culture and Conservation*, 200–203.
185 For such an unidentified continuity, see, for example, ICOMOS evaluation for the nomination of the World Heritage property, "Galle," Sri Lanka, No 451, July 1988.
186 ICOMOS evaluation for the nomination of the World Heritage property, "Sana'a," Yemen, No 385, April 1986.
187 Operational Guidelines, January 1987, Paragraphs 23–31.
188 ICOMOS evaluation for the nomination of the World Heritage property, "Timbuktu," Mali, No 119 Rev, May 1988; UNESCO, World Heritage Committee, Twelfth session (Brasilia, 5–9 December 1988), Report, SC-88/CONF.001/13, Paris, 23 December 1988, p. 17, http://whc.unesco.org/archive/1988/sc-88-conf001-13_e.pdf.
189 ICOMOS evaluation for the nomination of the World Heritage property, "Istanbul," Turkey, No 356, July 1985. For an opposing approach to using the World Heritage inscription as a means of enhancing wider urban protection, however, see ICOMOS evaluation for the nomination of the World Heritage property, "Aleppo," Syrian Arab Republic, No 21, April 1986.
190 ICOMOS evaluation for the nomination of the World Heritage property, "Lübeck," Germany, No 272, May 1983.

191 UNESCO, Bureau of the World Heritage Committee, Seventh Session (Paris, 27–30 June 1983), Report of the Rapporteur, CLT-83/CONF.021/8, Paris, 1 August 1983, p. 9, http://whc.unesco.org/archive/1983/clt-83-conf021-8e.pdf.

192 UNESCO, World Heritage Committee, Seventh Session (Florence, Italy, 5–9 December 1983), Report of the Rapporteur, C/83/CONF.009/8.5, Paris, January 1984, http://whc.unesco.org/archive/1983/sc-83-conf009-8e.pdf.

193 ICOMOS evaluation for the nomination of the World Heritage property, "Lübeck," Germany, No 272rev, April 1987.

194 Labadi, "World Heritage, authenticity."

195 See, for example, ICOMOS evaluation for the nomination of the World Heritage property, "Québec," Canada, No 300, July 1985.

196 For the term "complex authenticity" see von Droste and Bertilsson, "Authenticity and World Heritage," 8.

197 ICOMOS evaluation for the nomination of the World Heritage property, "Bern," Switzerland, No 267, June 1983.

198 ICOMOS evaluation for the nomination of the World Heritage property, "Bern," Switzerland, No 267, September 1983.

199 ICOMOS evaluation for the nomination of the World Heritage property, "Puebla," Mexico, No 416, April 1987; ICOMOS evaluation for the nomination of the World Heritage property, "Puebla," Mexico, No 416, October 1987.

200 ICOMOS evaluation for the nomination of the World Heritage property, "Cuenca," Ecuador, No 863, September 1999.

201 ICOMOS evaluation for the nomination of the World Heritage property, "Willemstad," Netherlands, No 819, September 1997.

202 ICOMOS evaluation for the nomination of the World Heritage property, "Historical City of Jeddah," Saudi Arabia, No 1361, 10 March 2011. UNESCO, World Heritage Committee, 35th Session (UNESCO, June 2011), ICOMOS Evaluations of Nominations of Cultural and Mixed Properties, p. 85–95, http://whc.unesco.org/archive/2011/whc11-35com-inf.8B1e.pdf; ICOMOS evaluation for the nomination of the World Heritage property, "Old City and Ramparts of Alanya," Turkey, No 1354, 10 March 2011. UNESCO, World Heritage Committee, 35th Session (UNESCO, June 2011), ICOMOS Evaluations of Nominations of Cultural and Mixed Properties, p. 307–316, http://whc.unesco.org/archive/2011/whc11-35com-inf.8B1e.pdf; ICOMOS evaluation for the nomination of the World Heritage property, "Jajce," Bosnia and Herzegovina No 1294, 10 March 2009. UNESCO, World Heritage Committee, 33rd Session (UNESCO, 22–30 June 2009, Seville), ICOMOS Evaluations of Cultural Properties, WHC-09/33.COM/INF.8B1, p. 112–120, http://whc.unesco.org/archive/2009/whc09-33com-inf8B1e.pdf.

203 ICOMOS evaluation for the nomination of the World Heritage property, "Graz," Austria, No 931, September 1999.

204 Pauline von Bonsdorff, *The Human Habitat: Aesthetic and Axiological Perspectives* (Lahti: International Institute of Applied Aesthetics, 1998), 125; Elisa El Harouny, *Historiallinen puukaupunki suojelukohteena ja elinympäristönä: esimerkkeinä Vanha Porvoo ja Vanha Raahe* (Oulu: Oulun yliopisto, 2008), 20.

205 ICOMOS evaluation for the nomination of the World Heritage property, "Panamá," Panama, No 790, September 1997; See also ICOMOS evaluation for the nomination of the World Heritage property, "Mauritanian towns," Mauritania, No 750, October 1996.

206 ICOMOS evaluation for the nomination of the World Heritage property, "Le Havre," France, No 1181, April 2005.

207 ICOMOS evaluation for the nomination of the World Heritage property, "Tel Aviv," Israel, No 1096, March 2003.

208 El Harouny, "Historiallinen puukaupunki," 269.

209 Ioannis Poulios, "Moving beyond a values-based approach," 181.

210 Marc Laenen, "Authenticity in relation to development," 353.

211 ICOMOS evaluation for the nomination of the World Heritage property, "Tel Aviv," Israel, No 1096, March 2003.
212 Green, *Managing Laponia*, 86.
213 See, for example, van Oers, "Managing cities," 7.
214 ICOMOS evaluation for the nomination of the World Heritage property, "San Cristóbal de la Laguna," Spain, No 929, September 1999.
215 ICOMOS evaluation for the nomination of the World Heritage property, "Vienna," Austria, No 1033, September 2001.
216 See also Vahtikari, "Historic cities."
217 von Droste and Bertilsson, "Authenticity and World Heritage," 14.
218 See also Labadi, "World Heritage, authenticity."
219 ICOMOS evaluation for the nomination of the World Heritage property, "Karlskrona," Sweden, No 871, October 1998; ICOMOS evaluation for the nomination of the World Heritage property, "Portovenere / Cinque Terre," Italy, No 826, September 1997.
220 ICOMOS evaluation for the nomination of the World Heritage property, "La Chaux-de-Fonds/Le Locle," Switzerland, No 1302, 10 March 2009. See also ICOMOS evaluation for the nomination of the World Heritage property, "The Canal Area of Amsterdam," Netherlands, No 1349, 17 March 2010; ICOMOS evaluation for the nomination of the World Heritage property, "Old City and Ramparts of Alanya," Turkey, No 1354, 10 March 2011. UNESCO, World Heritage Committee, 35th Session (UNESCO, June 2011), ICOMOS Evaluations of Nominations of Cultural and Mixed Properties, p. 307–316, http://whc.unesco.org/archive/2011/whc11-35com-inf.8B1e.pdf.

6   Outstanding universal value and the local narrative of place and heritage

Old Rauma

Until now I have explored the construction of outstanding universal value for cities as part of international and transnational heritage discourses and valuation practices. The story of a World Heritage city, however, does not end at the point of designation. Hence, in this chapter the focus will be shifted to the local level, to a city of medieval origin, Rauma, located on the west coast of Finland (Figure 6.1). Traditionally a port and a mercantile town, Rauma became industrialized beginning in the early twentieth century. Post-war industrialization then led to population growth in the context of this small community of a few thousand people, reaching 20,000 inhabitants by 1960.[1] Today Rauma has 40,000 inhabitants. Old Rauma, an area that was surrounded by a toll fence until the early nineteenth century (Figure 6.2), was inscribed onto the World Heritage List in 1991 as one of the first two Finnish sites.

The consequences of a World Heritage listing vary from city to city, depending on the particular context – historical, political, economic, social and legal – of each individual place. The most often reported impacts of the status are an increase in international and/or domestic tourism, improved heritage management, introduction of new instruments for protection, and more governmental, regional and/or private funding. Other consequences include altered uses of the site, gentrification, over-restoration, and changed representation of the place, or, in a more positive light, a greater contribution to a stronger local identity.[2] The listing of one place as World Heritage may also influence the non-listed locations in terms of conservation, visitation and the politics of heritage.[3] Moreover, the World Heritage inscription may also reproduce national heritage hierarchies, as one site is valorized over other sites with nationally similar profiles.

In this research the local consequences of the World Heritage listing will be addressed from a slightly different angle, that is from the point of view of the operation and (re)articulation of the notion of outstanding universal value within the local context. The focus will be placed on the views of different groups having a stake in Old Rauma as heritage, World Heritage and as a place of everyday living, and on their negotiation over meanings attributed to the World Heritage site. These groups include the city government, local conservation authorities,

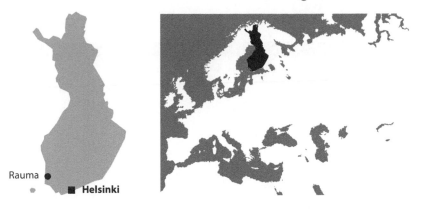

Figure 6.1 Rauma in Finland.

the National Board of Antiquities, Old Rauma shopkeepers and Old Rauma residents. This chapter thus begins with a discussion concerning the 'pre-history' of Old Rauma as heritage and a historic city, extending from the turn of the nineteenth and twentieth centuries up to the 1980s. The purpose of this discussion is not so much to give a detailed account of Old Rauma's conservation history, but rather to provide a solid basis for examining the local understanding of outstanding universal value in relation to various meanings and values associated with Old Rauma at earlier times: as local and national heritage, and, most importantly, as a "lively wooden town." The chapter then examines the establishment of World Heritage value for Old Rauma by comparing the value statement articulated by the Finnish State Party with the one produced by ICOMOS. The remaining part of the chapter deals with the operation of outstanding universal value locally after 1991, with a particular focus on the debate concerning the construction of a shopping center in the World Heritage buffer zone of Old Rauma in 2003–2006, something which finds parallels in many other World Heritage–designated cities.

My approach to the shopping center debate is a process-oriented one. I have followed this specific issue from beginning to end, exploring how different groups negotiate their standpoints during the process.[4] To echo Michael James Miller, I believe that it is especially in moments of change and threat to a local environment that the various groups involved give voice to their beliefs and values.[5] It was in the statements by the above-identified groups during the shopping center debate that the issues of World Heritage and outstanding universal value became articulated. The arenas in which the various groups presented their arguments included the local media, the planning process (plans, appeals, replies) and the court processes (appeals and pleas). A central theme addressed in this chapter is the relationship between local, national and international levels in the definition of a World Heritage city, and, in particular, how and to what extent the universal concept has shaped lower-level heritage meanings.

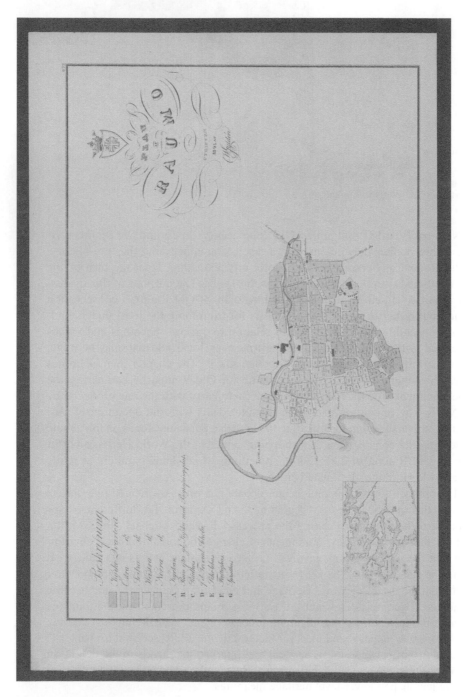

Figure 6.2 Rauma as portrayed by C. W. Gylden in 1841.

Source: The National Library of Finland.

Making Old Rauma: Different senses of heritage

"Cherishing the spirit of the old town"

It is possible to distinguish roughly five phases with regard to Old Rauma's history as heritage, following the general European trend.[6] During the first period of discovery, from the turn of the nineteenth and twentieth century until the First World War, Old Rauma was named and produced as a historic city in relation to the growing and industrializing 'new town,' which physically extended beyond its old borders. The need for protecting Old Rauma – with its (partially) medieval street pattern and old wooden buildings dating from different periods – was both a locally motivated and exterior elite concern, underpinned by nostalgic and nationalist discourses. The second distinct phase was the interwar period: while locally valued, in the context of the growing concern among the Finnish art historians for the physical fabric and exterior aesthetics of historical cities, Old Rauma was considered a less-suitable representative of an iconic Finnish historic city, resulting in a somewhat ambiguous position within the national heritage hierarchy. This was because of the Neo-Renaissance-style changes and remodeling carried out in most Old Rauma houses in the late-nineteenth century, which were stylistically not in conformity with the architecture of classicism favored at the time.[7]

Third, as might be anticipated, the post-war period, up until the late-1960s, saw a partial marginalization and redefinition of the heritage meanings attributed to Old Rauma from the point of view of modernism in urban planning and architecture. The established perceptions of what constituted the old town were challenged, as the historic city was reduced to the level of abstract variables on the order of "scale," "panorama" and "overall attractive nature," and to the slogan "cherishing the spirit of the old town."[8] Thus, in the 1960s, the historic city had a ritual rather than a pedagogical function as part of the local memory: it was valued more as a place of public ceremony than for its authentic structures. In the early 1960s, the city government and certain property owners promoted a large-scale renewal of the central blocks of the old town area. As in many other places, these plans were rationalized in reference to the needs of modern business, land use efficiency in the city center, and improvement of sanitary and structural conditions of houses considered inferior.[9] It is true that in comparison to modern building, the standard of living in Old Rauma during the 1960s was modest and the area's social structure, with many small households and a relatively low income level, was different from the equivalent elsewhere in the city.[10] Still, as I have pointed out elsewhere, there is no need to exaggerate the derelict state of the area or its renewal as a broadly shared narrative among Old Rauma residents. There was important social investment in the area, even in the 1960s.[11] The meanings attached to Old Rauma in the 1960s can thus best be described as ambivalent.[12] While the physical renewal of the area was considered a necessity, historical and aesthetic values were equally underlined. This was apparent, for example, in the summary of the 1964 architectural competition for the city center, which understood at least partial physical renewal of the area to be inevitable, but which nevertheless took as a starting point the old town, "a well-preserved body of wooden houses, whose town plan was of medieval origin."[13]

The fourth, and in many ways the key, period in the history of Old Rauma as heritage dates from the late 1960s to the early 1970s. During that period two important developments took place throughout Europe. Firstly, the modern approach to conservation and restoration was adopted, shifting these practices "from an artistic to a critical sphere."[14] Secondly, the period witnessed a growing public consciousness of and public protest in favor of cultural heritage and other environmental issues. Hence, heritage work became more pluralistic and democratic as various groups 'from below' started to join the conservation professionals in support of protection.[15] Moreover, in the conservation and rehabilitation of cities, social criteria became linked with architectural and historical values.[16] Also in the context of Old Rauma both these developments were introduced at the turn of the 1960s and 1970s. The so-called framework plan for Old Rauma, preparation of which had started in 1964 after the architectural competition, still promoted the objective of a physical renewal of the area, and treated heritage mostly in terms of scale and spirit. When completed in 1972, after lengthy preparation,[17] the framework plan was regarded as outdated from the perspective of the protective agenda supported by many experts and citizens alike.[18] In the meantime, the National Board of Antiquities had adjusted its earlier, 1966–1967 inventory of Old Rauma towards a more inclusive, and more scientific, treatment of the old building stock.[19] Some solutions proposed in the plan, especially the large-scale renewal of the business district and the proposal to build a new street across the western part of Old Rauma, were openly criticized by citizens when introduced to them in March 1973.[20]

The relatively quick transition from the discourse and practice of physical renewal to the principles of preservation of the historical fabric is also visible when going through the articles about Old Rauma published in the local newspaper. While the *Länsi-Suomi* newspaper in early 1973 still praised the framework plan as an optimal solution in terms of the future development of the area,[21] the articles published during the following year presented Old Rauma, in its existing state, as a model for future urban planning, and discussed the economic and technical requisites for protection.[22] In 1975 the newspaper was already publishing a highly romanticized description of Old Rauma titled "The Old Rauma song":

> It pities the geometric boredom of apartment buildings and unimaginative living in boxes. The song of Old Rauma is a tale about smoke from the saunas, the scent of clean laundry in the dusk of Saturdays, children's play around their home gates, cups of hot coffee at the town square, everyday work, joyful roses on the side of the hill. It is a tale about life rather than mere living.[23]

A process of planning was reinstituted and the town plan amendment for Old Rauma, based on broad protection of existing buildings across the entire old-town area, gained legal status in 1981 (Figure 6.3). In that plan, 60 percent of the buildings (343) were marked for immediate protection (not to be demolished

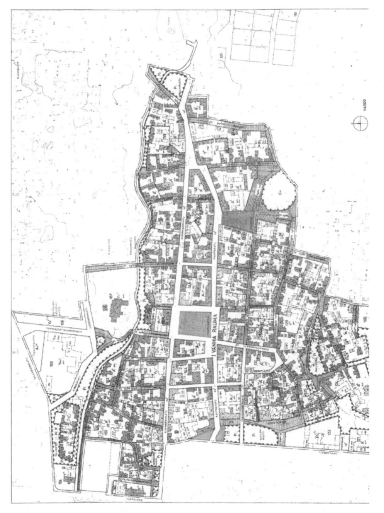

Figure 6.3 The Old Rauma Conservation Town Plan from 1981. S stands for buildings marked for immediate protection, and H for buildings recommended for conservation but considered potentially replaceable. White areas mark the building stock considered recommendable for rebuilding.

Source: City of Rauma.

without compelling reason) and another 10 percent (65) were recommended for conservation but considered potentially replaceable by new "environmentally-suitable" buildings. The rest of the building stock was considered available and even recommendable for demolition and rebuilding.[24] This up-to-datedness of the town plan on the national heritage agenda was one reason why the Finnish government decided to nominate Old Rauma for inscription on the World Heritage List in 1990, after having ratified the World Heritage Convention in 1987.[25] The inscription of Old Rauma on the World Heritage List in 1991 started the fifth and final phase in Old Rauma's history up until then as heritage.

Key value: "lively wooden town"

In Rauma, it was acknowledged early on that Old Rauma was something that distinguished the city from many other small cities in Finland. This was seen as due to its historical associations and townscape but also because Old Rauma continued to form the commercial and public center of the city, something that was no longer so common among Finnish cities, even in the 1930s:

> The city of Rauma combines, in our country, in a unique way, a historic town plan originating from medieval times and a technologically advanced modern city's vibrant entrepreneurial spirit. In a miraculous way the old historic part of the town, with its narrow winding streets and changing idyllic cityscapes has been spared destructive fires throughout its centuries. Equally peculiar is that the old town has, as the only exception in our country, managed to maintain its position as the commercial district of the city. This fact, with its many problems produces many challenges for contemporary city planners, but makes it all the more interesting.[26]

As can be read from the above statement by Rauma mayor Alfons Tammivaara, during the interwar years, the dual function and meaning of Old Rauma as heritage, on the one hand, and as a commercial city center, on the other, were not considered in any way contradictory. Liveliness became a central meaning attached to Old Rauma and an element that was thereafter repeatedly reproduced. For instance, the jury of the 1964 architectural competition for the city center actively promoted the concept of liveliness by underlining that relocating the city center entirely outside Old Rauma, as had been proposed in some competition entries, was unrealistic, since businesses also had a long tradition of being situated in the area.[27]

Similarly, one key objective of the 1981 town plan amendment was to support the functional diversity of Old Rauma and to provide requisites for continuous life and development in contrast to the then criticized ideology of creating a city museum. Even though the established division between the business zone in the western part of Old Rauma and the residential zones was maintained in the plan, an effort was also made to mix different functions. The plan made several concessions to the viability of the business area and the needs of shopkeepers. On average, more building rights and fewer restrictions were introduced for commercial buildings in comparison to residential buildings. Many of the buildings rated as second-class in the plan, recommended for conservation but considered as potentially replaceable by new "environmentally-suitable" buildings, were situated along the main business streets of Kauppakatu and Kuninkaankatu. Moreover, despite the year-long debate during the 1970s concerning the creation of pedestrian streets in Old Rauma, the plan allowed vehicle traffic and car parking to continue in the central commercial blocks, an objective strongly lobbied for by the shopkeepers in the area (Figure 6.4).[28]

Despite these concessions to the shopkeepers, many of them found it difficult to accept the town planning amendment's guiding conservationist principles. During the different phases of the planning process a large group of shopkeepers joined together to launch appeals. In general, the shopkeepers were afraid that detailed protection would freeze Old Rauma's role as part of the city center and its ability to change in the future. In the shopkeepers' view, the 1981 Town Plan Amendment promoted the ideal of reconstituting the situation which had prevailed in the late nineteenth century, both in material terms and in terms of requiring the revival of past forms of commerce. Two metaphors with opposite meanings used frequently in the shopkeepers' argumentation in the 1970s and 1980s were "liveliness" and "museumification," both already familiar images from the debates during the 1960s – losing the former would eventually lead to the latter. In the context of the drafting of the town planning amendment the metaphor of museumification was redefined. It was no longer used to argue for large-scale physical renewal, as was the case in the 1960s, but for allowing change despite the protected status.[29]

Figure 6.4 Old Rauma street, Kuninkaankatu, in 1974.

Source: Rauma City Museum.

The shopkeepers feared that protection would mean loss of control of Old Rauma as an economic resource.[30] In many shopkeepers' arguments, allowing vehicular traffic in Old Rauma represented a particularly important symbol of its liveliness: restricting car traffic would lead to the eventual "death" of the area.[31] Illustrative of these sentiments was the local newspaper's cartoon picturing a conversation between two women on one of the curving streets of Old Rauma: "Oh dear, they are making museum pieces of us," said the one. "Yes, and soon they will want us to wear long gowns too," replied the other.[32]

For many Old Rauma residents the limitations on building rights were easier to accept, since these objectives did not stand in stark contrast to the use of Old Rauma as a residential area. Also, many people who were positively disposed towards the valuation of old wooden towns had moved into the area during the 1970s. In 1974 the inhabitants founded the Old Rauma Society (*Vanha Rauma yhdistys*) for the protection of the area's architectural and historical values. Both these developments reasserted the voice of residents in negotiations concerning Old Rauma. In addition, while the local conservation and planning authorities were committed to maintaining the functional diversity of Old Rauma, the predominant heritage discourse tended, at least implicitly, to support the residential use of the area, precisely because of the problem-free nature of this use.

Even though the approval of the Town Plan Amendment in 1981 consolidated the protected status of Old Rauma through laying down general principles for future treatment of the area, the early implementation of the plan in the 1980s was not devoid of ambiguities and controversies. On the one hand, these could be seen in the form of a few demolitions and some major remodeling of buildings, all cases where the views of authenticity and historical value of conservation authorities and property owners (mainly shopkeepers) stood in stark contrast.[33] For many shopkeepers authenticity was not compromised when an old building was replaced by a new one. As summarized by the local newspaper: "Among the entrepreneurs the renewal of the buildings is seen as the preservation of the historical image of Old Rauma and as securing the continuation of the district as a vibrant and unique place for commerce. It is felt that this not only has wide significance for the economy and services and thus wide significance for the city and its inhabitants' wellbeing, but also that it has permanent cultural and historical value."[34] On the other hand, ambiguity, even though different in kind, characterized the Town Plan Amendment's guidelines as well as certain restoration projects. Old Rauma was viewed not only in terms of conservation measures but also in terms of "corrective measures."[35] For example, all of the (few) larger-scale industrial and warehouse buildings (a wool-spinning mill and dye house, a sausage factory), built of stone, were classified as environmentally unsuitable for Old Rauma in the name of townscape unity. Consequently it was proposed that small-scale wooden residential buildings be built on the relevant plots to replace them,[36] thus reinforcing the existing wooden town typology.

The above discussion is illustrative of diverging and coincident meanings concerning the historic city, authenticity and change in the context of Old Rauma over the decades. When a site becomes designated as World Heritage, another meaning is attached to the place and the heritage. However, when implemented locally, international proclamations such as the World Heritage Convention and the concept of outstanding universal value are inseparably bound to the historical context of a particular place, and the earlier meaning-making related to it. In other words, the existing relations in the local community, older issues and interests, and the earlier construction of place meanings all surface as part of the discussions and debates concerning World Heritage nomination and designation and later site management. It is against this background that we should explore the concept of outstanding universal value, and its re-articulation and operation in the local context. That will be the focus of the remainder of this chapter.

Outstanding universal value of Old Rauma: Two definitions

According to the overall tendency in the early phase implementation of the World Heritage Convention, the city of Rauma pursued no conscious strategy for gaining World Heritage status.[37] There was also a total lack of consultation with Old Rauma homeowners, shopkeepers and inhabitants during the process.[38] As with many other sites nominated at that time, the Finnish nomination dossier concerning Old Rauma was concise, counting less than 50 pages including all annexes (nowadays this documentation often numbers 500 pages). I will now discuss the national nomination dossier from the perspective of the understanding of heritage and outstanding universal value articulated in it. Three features in particular should be pointed out. First, like the Town Plan Amendment a decade earlier, the nomination file stayed congruent with the wooden town typology, placing the outstanding universal value within the Nordic context. The Nordic countries had jointly harmonized respective World Heritage nominations in three meetings initiated by ICOMOS; in these discussions the 'wooden town' had appeared as one of the self-evident categories to be represented on the World Heritage List.[39] Consequently, the Finnish nomination concerning Old Rauma was firmly founded when it was possible to state that Old Rauma represented "one of the 150 wooden towns left in Norway, Sweden and Finland," and that timber-house towns constituted "a clearly definable phenomenon" that had played "a decisive role in European urbanization."[40]

Secondly, while the nomination suggested chronological continuity by referring to different centuries since the establishment of Rauma in the fifteenth, the period that was highlighted as the major period of significance was the late-nineteenth century, which had been marked by the thriving sailing-ship trade, an economic boom and the consequent remodeling of the city center. The late-nineteenth century remodeling, while seen as an intrusion by the interwar preservationists, had become highly valued by the early 1990s. In fact, Neo-Renaissance was the only architectural style mentioned in Old Rauma's World Heritage nomination file (Figure 6.5).

Thirdly, and most importantly, even though the outstanding universal value and authenticity of Old Rauma were based on architectural unity, the well-preserved historical building stock and a street network that could be traced back to the

Figure 6.5 One of the most imposing Neo-Renaissance buildings, *Marela* in Old Rauma.
Photograph by Touko Berry.

Middle Ages, they were also considered to be due to "a lively community with various services, residential buildings and shops."[41] Thus, liveliness, long in the making as a key determinant of the identity of the place, was integrated into Old Rauma's outstanding universal value proposed by the State Party. Important constituents of this liveliness were seen to be the 180 shops in the area, which formed a center of retail commerce in the city.

During its annual meeting held in Carthage in December 1991, the World Heritage Committee inscribed Old Rauma on the World Heritage List.[42] As was common practice at the time, when making its decision to inscribe Old Rauma, the Committee made no further comments on the issue, which implies that it fully accepted the evaluation and recommendation that ICOMOS had decided on a few months earlier. The outstanding universal value of Old Rauma had been evaluated by ICOMOS in May 1991, establishing it, as with the Committee later on, in reference to criteria iv and v for inscription of cultural sites:

> Criterion V. Rauma is an outstanding example of an old Nordic city constructed in wood, a veritable conservatory of traditional settlements in this part of Europe.
> Criterion IV. Consequently, this city is typical of the architecture and urbanism of old North-European cities and is one of the most beautiful and extensive of all those preserved thus far.[43]

Nowadays, when inscribing a new site on the World Heritage List, the Committee adopts an official statement of outstanding universal value, which is to form the basis for future management of the site. With the early World Heritage sites, the ICOMOS evaluation factually made the statement of outstanding universal value. This is also the case with Old Rauma; the use and wording of the criteria, as well as the themes and values brought up in the broader evaluation text, can be understood as such a statement.[44] The evaluation document highlighted Old Rauma's typicality and representativeness within the North European context. It also drew attention to the following: wood as a building material; the "ancient aspect" that Old Rauma had preserved "despite some changes made in the nineteenth century"; the aesthetic value ("one of the most beautiful"); the historical and age value ("old Nordic city," "conservatory of traditional settlements"); the extensiveness of the heritage resource ("one of the most extensive of all those preserved thus far"); and the confirmed conservation status of the area ("half of the buildings in the old city already restored"). Temporal references were made to the medieval period, the late seventeenth century, and the nineteenth century, but the latter, interestingly enough, was mentioned only as part of the imprecise statement implying that only very few changes had been made during the nineteenth century. Here the ICOMOS evaluation clearly departed from the national nomination dossier, and proposed a more traditional view very much in accord with the overall ICOMOS evaluation practice of the time. It was also considered important to depict Old Rauma in relation to its few monumental public buildings (the Franciscan church, the ruins of another medieval church and the City Hall) all built of stone, even though the ICOMOS evaluation referred to vernacular architectural heritage as Old Rauma's "great wealth."[45]

One further issue that needs to be pointed out concerning the ICOMOS evaluation document is the use of criterion v, and ICOMOS' notion of Old Rauma as "a veritable conservatory of traditional settlements." As was pointed out in Chapter 5, ICOMOS, in reference to most cities, used this criterion in the meaning of anti-modern and stagnant. This was also the case with Old Rauma, and as a result ICOMOS ascribed a meaning to Old Rauma foreign to the local community. Although this meaning was not contested by locals at the moment of inscription, it should be borne in mind that there was very little possibility for voicing such a critical opinion, since the locals were not informed or consulted about any of the value statements at the time of the inscription. All in all, on the basis of the above outlined trajectory of Old Rauma into the place of heritage, as well as my interviews with the local stakeholders, all referring to liveliness as a key aspect of Old Rauma as opposed to attributes of "stagnation," "a stage" or "a museum town,"[46] it is fair to assume that they – even the conservation authorities, let alone the shopkeepers – had never ascribed a meaning to Old Rauma as "a veritable conservatory of traditional settlements." While the ICOMOS evaluation reproduced some of the statements made in the national nomination document, the overall emphasis in the Finnish nomination dossier had nevertheless been different. ICOMOS chose not to reproduce the idea of a lively community.

The ICOMOS official wording of outstanding universal value was not distributed locally after the designation. Instead, the definition given in the national nomination dossier was interpreted as the official statement of outstanding universal value by local stakeholders and also by the National Board of Antiquities.[47] Since it is this national definition that has been central to the discussions concerning World Heritage in the context of Old Rauma, two things should be pointed out here. Firstly, this dismissal of the international definition shows that it has not always been clear within the World Heritage system as to who and what makes the statement of outstanding universal value, especially at the early stage of the implementation of the Convention. While the lower-level statements of outstanding universal value have not always been integrated into the international-level statements, the international-level statements have equally failed to translate to the local and national levels. Secondly, the centrality of the local and national definition in the context of World Heritage–designated Old Rauma shows the endurance of local place meanings. Therefore, for Old Rauma, outstanding universal value, even though a concept fully initiated from above and a process lacking any genuine bottom-up identification of values,[48] became firmly rooted in the earlier local understanding of the area.

"Lively wooden town," but to what extent?

After Old Rauma gained World Heritage status, the earlier discussions about conservation and liveliness picked up thematically from where they had left off, but this time in a changed situation. The first change had to do with the designation itself, for which the various local interest groups had different but equally high expectations. The second change related to the changed economic situation in the city, and throughout Finland. During the economic boom of the late 1980s the demand for commercial space in the city center had increased, sometimes leading to major remodeling of the Old Rauma properties. But when the boom ended and the economic recession began in the early 1990s, it was suddenly difficult to find tenants for all the remodeled business premises in the area. This caused the area to deteriorate and become unsightly, with fewer shops operating and old specialty stores having changed into bars or flea markets.[49]

All local stakeholders in the early 1990s were of a common opinion about the somewhat derelict state that Old Rauma was in. In contrast, their solutions, and their understanding of how the World Heritage status would fit into the overall picture were different. All groups began to use outstanding universal value as an argument to support their own visions for Old Rauma. For their part, most shopkeepers viewed the new status as a means to revive a city center that had been hit hard by the recession. What was at stake again, as articulated by the local newspaper, was the liveliness of Old Rauma: "The foundation that Old Rauma has built its fame upon is crumbling at a fast pace. As long as this continues, it cannot be regarded as a center of lively commerce or as a concentrated and broad collection of specialty stores, as it used to be."[50] The reference to Old Rauma's "fame" and the possibility of losing that fame if the commercial prerequisites were weakened, established a discursive link between the World Heritage status and the shopkeepers' earlier, and in the early 1990s, accelerated concern for the commercial liveliness of the area.[51]

During succeeding years, some Old Rauma shopkeepers were of the opinion that Old Rauma would not have been granted World Heritage status in the first place without their presence and the notion of a "lively community" in the World Heritage bid. Since the ICOMOS evaluation document and its use of qualification criteria making reference to "a conservatory of traditional settlements" fell short of such a direct link between the commercial life and World Heritage status, this position of the shopkeepers seems overstated. The status of Old Rauma, as representative of 150 Nordic wooden towns and as confirmed in negotiations between the Nordic countries, would probably not have been affected one way or another in the minds of UNESCO and ICOMOS representatives if the nomination had focused solely on architectural and historical qualities without any mention of the lively community. However, as we have seen, the definition of outstanding universal value that was locally adopted was not the one articulated by the international stakeholders but the one outlined in the national nomination dossier, something which found resonance among local stakeholders and was in conformity with the earlier decades' construction of place and heritage meanings. Against this background it is understandable that for Old Rauma shopkeepers World Heritage was mainly about protecting liveliness, which meant maintaining a thriving business in the area. Any attempt to weaken the commercial prerequisites in Old Rauma would threaten the existence of Old Rauma as a World Heritage site. Thus, World Heritage, even though a conservation initiative, became interpreted by some as an impetus for change.[52] Moreover, from the shopkeepers' viewpoint their ability to work and carry on successful business in the area, and in essence even the shopkeepers themselves, became an aspect of the heritage that needed to be protected as World Heritage.

The concern for the liveliness and functional continuity of Old Rauma voiced by shopkeepers continued to be motivated by economic interests and a concern for their own livelihood, a concern that the city government also shared.[53] Also underlying the shopkeepers' views was a different understanding of what constitutes heritage, a functional rather than an aesthetic and material attitude towards heritage. For many of them Old Rauma's heritage was primarily bound up with Rauma's maritime and merchant past, with the professional heritage of, for example, goldsmiths or furriers, or with the history and heritage of their own family, and their family's business.[54] Accordingly, the sense of heritage of Old Rauma was understood more in terms of its intangible aspects, an attitude also reflected in the conception of authenticity, considered not only as an aspect of the past but also as an aspect related to the present and the future:

In my opinion it is very dangerous when a small group of people tells us what is authentic and what is not. In a way, the development is frozen in that moment, for instance in the early twentieth century, even though there have been things authentic also before and after. There is authenticity also today.[55]

Liveliness also remained part of the conservationists' language. The shopkeepers' argument that shops were keeping the townscape alive and justifying Old Rauma as a World Heritage site acquired resonance amongst the national-level heritage authorities, although shops were also considered as problematic from the conservation point of view.[56] At the local level, certain reservations continued to be expressed regarding the meaning of liveliness and its scope, as had been the case during the previous decade. Even though local conservation authorities shared a concern about the derelict state of Old Rauma and the objective of functional diversity, they did not share the shopkeepers' view about the all-sacred nature of Old Rauma's commerce:

> One criterion for the liveliness was that this is a versatile area. A versatile nature is a good thing, but not to just any extent. It is not a great sin if old shops are turned into apartments in a place where only a shady pub can maintain business. One must not underestimate people residing in the area, it is equally important. [...] There are many kinds of liveliness. Both are needed, residences and businesses, and no exact critical limits or proportions have been set up between these two functions. If the area is dominantly about shops, it is vibrant during the days, but dead in the evenings.[57]
>
> A commercial center might often be a bigger threat to World Heritage than a supporting factor. It usually means that there has been a lot of change in the building stock. In the case of Rauma neither factor can be separated from the World Heritage status, but, after all, this is a historical and an architectural monument, so even without the commercial center it could be a part of World Heritage.[58]

In Finland, since the 1950s, urban renewal and architectural discourses have connected liveliness to people's presence and ability to sojourn in the city centers, as well as to consumption and commercial interests.[59] However, when the dynamic nature of a city is based solely on economic aspects, the ultimate result for many other areas of life, may be less regenerating, leaving a "dead city." From its functions the city may appear very lively on the surface but still cause feelings of otherness for its citizens.[60] The Old Rauma conservation authorities shared this understanding of a more diversely defined liveliness. And very much in line with the arguments of the 1980s, they considered the residential use of the area the most compatible with their conservation agenda, which was now supported by the notion of outstanding universal value. Local conservationists saw World Heritage as a means of developing Old Rauma's conservation in the direction of a more detailed approach and to end certain commercial functions in the area, considered unsuitable and problematic, because they, more than with residential use, required concessions from the point of view of material authenticity. Thus, in the 1990s,

World Heritage status further supported the joining together of conservation and residential meanings of Old Rauma. The uneasiness with certain commercial uses, together with the difficulty of drawing the line between acceptable and unacceptable functions, becomes apparent in the following extract from an interview with a local conservation authority: "There were hardware stores, which started to be too big. When has the limit been reached so that you simply need to relocate? Now this has by and large taken place; they have moved elsewhere, outside Old Rauma. A smaller hardware store, though, would make a nice addition to the area."[61]

In retrospect, the conservationists' agenda was successful. With regard to Old Rauma, the 1990s clearly meant a period of naturalizing the meaning of the historic city based on the preservation of authentic historic fabric both discursively and at the level of conservation practice. This was the case, even though no new conservation regulations or management structures were put in place following the World Heritage inscription. The 1981 Town Plan Amendment remained the main legal basis for the area's treatment. It is difficult to assess how much this consolidation of meaning was due to the World Heritage status, and how much of it resulted from the overall discursive change in urban heritage conservation in Finland, which was marked by ever more detailed conservation requirements. Perhaps it is fair to say that in Old Rauma's case, the World Heritage designation worked to strengthen the heritage-oriented discourse, which was also topical for other reasons. In the process, the conservation authorities' symbolic ownership of Old Rauma increased.

Discussions with local stakeholder groups in Old Rauma reveal that World Heritage shaped general attitudes positively towards the area and its protection, even for many who originally opposed these ideas.[62] By the year 2003, the protected Old Rauma had become such a shared narrative that one interviewee, a long-term municipal politician, who had favored the renewal of the old town in the 1960s, remembered being in favor of conservation.[63] As in many other places, international recognition and outside visitors contributed positively to the consciousness of the inhabitants of Rauma about their own heritage.[64] When viewed from outsiders' eyes, one's own heritage may reflect its value and uniqueness in a novel way. This was the case with one shopkeeper interviewee who felt joy and pride when asked by two German couples to be guided around Old Rauma.[65]

The consolidation of heritage-oriented discourse with regard to Old Rauma in the 1990s also becomes evident in an analysis of the decisions made between 1982 and 2003 by the Old Rauma Committee (*Vanhan Rauman erityiselin*). The Old Rauma Committee is an advisory organ within the city of Rauma's building inspection organization which issues statements concerning all significant building and renovation projects in Old Rauma. The Committee has both expert and lay representatives. While during the 1980s the Committee mainly discussed individual large-scale renovations, since the 1990s it has concentrated more on either detailed planning or general level discussions, such as the ones concerning authenticity and the historicizing treatment of façades.[66] From 1982 to 2003 we find a decreasing trend in the number of matters discussed by the Committee. In 1982–1991 the Old Rauma Committee issued, on average, 48 statements per year followed by, on average, 27 yearly statements between the years 1992 and 2003. At the same time the share of negative decisions issued by the Committee also decreased slightly. Between 1982 and 1991 the Committee examined 91 cases and issued negative decisions in 16 percent of them; the rejection rate was down to 12 percent in the 1992–2003 period. Even though not dramatic, these figures suggest that during the latter period, the property owners had been somewhat more aware of the limits on acceptable and unacceptable change when drawing up their plans. Furthermore, some of the unsuitable proposals had been screened out before being brought to the Committee's table.[67]

An analysis of the statements by the Old Rauma Committee in the 1992–2003 period also reveals that World Heritage was referred to in only three percent of them, suggesting that World Heritage was not seen as particularly important to local reshaping of the place. Nevertheless, the reference to World Heritage was made in 13 percent of the most difficult and controversial cases rejected. One such rejection concerned a proposal to reconstruct a part of the old toll fence.[68] Clearly, a World Heritage reference, when used, facilitated "a weapon against the craziest ideas," [69] since poorly drafted plans could be rejected right away by referring to an upcoming periodic monitoring process. Moving back to the general discourse, it is worth noting that contrary to the public discourse in the 1980s when Old Rauma's protected status had been occasionally used as an argument for demolition in other parts of the city,[70] in the 1990s demands were voiced that Old Rauma's World Heritage status required protection of the old building stock in other parts of the town as well.[71]

All the above examples testify to the naturalizing of the historic-city meaning of Old Rauma subsequent to its obtaining World Heritage status. Even though the outsider agency of World Heritage contributed positively to the consciousness of the people in Rauma of their own heritage, some of the interviewees also associated World Heritage List membership with a negative point of reference in relation to the community's own heritage, and as having more meaning for outsiders than for the people actually living in Rauma.[72] Such statements show that it was difficult, at least for some, to truly associate Old Rauma with World Heritage. In the words of the local community members, UNESCO was often mistakenly called "UN" or "UNICEF." The abstractness of the World Heritage idea resulted in a situation in which it was unclear what the status really meant.

Moreover, while the conservation authorities adopted World Heritage as an argumentative device, even their use of it was selective. This could be seen for example in the late 1990s when the regional museum and conservation authority, Satakunta Museum, launched the EU-funded regional heritage project 'Hopes and Reality' (*Toiveet ja todellisuus*). Old Rauma was selected as one of the main objects of the project with the expectation that it would provide a means to obtain some additional funding for private homeowners in the area. However, it is striking that in the overall goals of the project, Old Rauma was treated almost like any other site in the region. Similarly, the extensive final report on the project hardly mentioned Old Rauma's World Heritage status. It may well be that it was the steering group's conscious decision not to position any one site hierarchically above the others. Nevertheless, it seems peculiar that when Old Rauma was discussed within the framework of the project, there was no elaboration on what specific conditions the World Heritage status would entail in terms of the management of the site. Even in those cases where the project openly criticized some of the established conservation practices employed in Old Rauma, there was no reference to the compatibility or incompatibility of these practices with regard to the defined out-standing universal value or the World Heritage Operational Guidelines.[73]

The criticism expressed by the steering group of the 'Hopes and Reality' pro-ject towards the established Old Rauma practices focused especially on certain exterior renovations,[74] an issue that was also discussed simultaneously by the Old Rauma Committee. The exteriors of many commercial and some residential buildings had been extensively remodeled between the 1950s and 1970s, espe-cially through installing large shop windows. By the end of the 1990s the ori-entation among shop and homeowners in Old Rauma had become reciprocal: to correct the 'mistakes' made in the post-war period, and to restore the earlier, often Neo-Renaissance appearance of the houses, also favored by local conservation authorities. The shop owners wanted to participate in the townscape improvement and common care for World Heritage. The expectations linked to tourism also played a role in their enthusiasm. Sometimes these correction measures involved selecting between several possible historical phases, or even approaching the point of historical invention by returning to a period that only existed in plans and drawings. As noted by one conservation authority, the turn of the nineteenth and twentieth century was considered a "safe" choice for restoration – "going further back in time would feel like falsification of history."[75] Going back to the turn of the nineteenth and twentieth centuries, however, was not regarded as presenting a similar kind of problem in relation to later periods. Here we find another set of converging views at work in the definition of Old Rauma as heritage in the 1990s, this time between those held by national and regional experts represented in the steering group (representatives of the National Board of Antiquities, the regional museum, academics) and those of local conservationists, who welcomed the façade reconstructions in the name of townscape unity.[76]

What was also at work in the local discourse, unintentionally and by no means uniquely or most flagrantly among World Heritage–designated cities,[77] was how to better fit the built urban structure into the image of a World Heritage city. Two levels may be discerned here: firstly, what was considered suitable, or unsuit-able, for World Heritage cities in general (for example an old appearance, no traffic), and secondly, how Old Rauma could present itself as a city with a sig-nificant late-nineteenth-century past, identified in the World Heritage documen-tation. Going back to one particular period in history, even if presenting itself as a "safe choice," is never unambiguous.[78] This concern was also voiced by some interviewees.[79] There is a fine line to be drawn with regard to the point of historical invention, since outstanding universal value easily serves to legitimize certain interpretations of the past, and certain visions of the future. These can be medieval images, seafaring and Neo-Renaissance pasts, or multicultural Jewish pasts overshadowing the rich and more recent industrial heritage in the public image of a town, as has been shown to be the case with the World Heritage city of Třebíč.[80] World Heritage not only labels 'fixed' heritage sites but participates in their 'making.' Still, it is not difficult to relate to the dilemma faced by local conservation authorities: who, among the shopkeepers, could be asked to preserve a historical appearance associated with the expansion of commerce in Rauma after World War II but generally considered unattractive by the local community today?[81]

During the 1990s World Heritage was – selectively – adopted as part of the local meaning related to Old Rauma. The notion of outstanding universal value was used to support various concepts concerning the conservation and the residential and commercial use of the area. The differing meanings associated with Old Rauma between shopkeepers and conservationists on the one hand, and between the national and local level heritage professionals on the other, could be best described as submerged tensions, as there was no overt conflict over Old Rauma and World Heritage during the 1990s. This situation changed, however, in the course of the new millennium. The next section will therefore focus on a moment of heritage dissonance, and examples of how outstanding universal value became articulated with the meaning of liveliness and other values, as part of the debate.

Protecting liveliness as outstanding universal value: Building a shopping center in the World Heritage buffer zone

At the time of Old Rauma's inclusion on the World Heritage List in 1991, no buffer zone for the World Heritage area was specified in the official nomination dossier, even though a map indicating the proposed World Heritage zone and a larger area surrounding it was annexed to the application by Finnish authorities. The Old Rauma example thus testifies to the practice of artificial and random delimitation of World Heritage areas and their buffer zones during the early implementation of the Convention, as discussed in Chapter 5. The beginning of the new millennium, however, experienced an accelerated national and local level concern for establishing a buffer zone around Old Rauma as a response to the considerations within the international heritage community.

The city of Rauma decided to define an official World Heritage buffer zone for Old Rauma within the context of the drafting of a City Centre Master Plan (Figure 6.6). The preparation of the plan started in the late 1990s and the plan was adopted by the City Council in August 2003.[82] The final delimitation for the buffer zone area was determined by local conservation authorities, together with the National Board of Antiquities. On the City Centre Master Plan, the buffer zone was marked with a note stating that "all plans involving the area and its future functions should take into consideration the special status of Old Rauma as a World Heritage site from the points of view of cultural history and townscape." Moreover, it was underlined that all significant projects involving the area should be submitted beforehand for consideration by the National Board of Antiquities.[83]

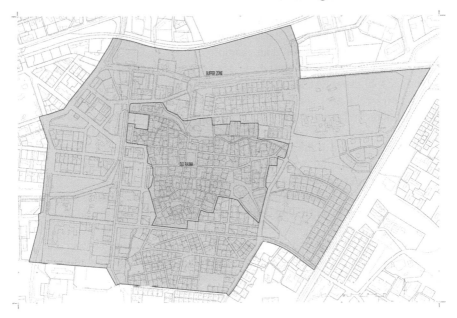

Figure 6.6 Map of Old Rauma buffer zone.

Source: City of Rauma.

This national decision was not officially submitted to the World Heritage Committee decision-making process until in 2009.[84] However, it was precisely within this nationally defined buffer zone, in an area located only 300 meters from Old Rauma on its northern side, where a major shopping center was controversially planned in 2003–2004 and built in 2006–2007. One could dismiss a discussion of this controversy outright simply because the buffer zone had not been officially approved by the World Heritage Committee at the time of the shopping center debate. These kinds of procedural failures testify to the insufficient knowledge on the part of states, Finland in this case, about the logic and rationale behind the definition of buffer zones within the World Heritage system. They even testify to a certain ambiguity within the World Heritage regime. According to the Finnish World Heritage coordinator in the National Board of Antiquities, Margaretha Ehrström, it was unclear in 2003, even for the World Heritage Centre experts, to what kind of procedures the buffer zone decisions should be exposed.[85] Nevertheless, the important point is that from 2003 onwards there was a common understanding among national and local stakeholders that a World Heritage buffer zone around Old Rauma had been established and that it had become operational.

Most of Old Rauma is surrounded by residential areas made up of single-family homes dating from the first half of the twentieth century. The area north of Old Rauma, however, occupies a different status.[86] In addition to one residential area with single-family homes from the 1950s, the area also includes another residential area built in the late 1970s and early 1980s consisting of higher apartment buildings – generally considered an example of less-successful urban planning in Rauma – a bus station, a filling station, a bakery, a drive-in hamburger restaurant, as well as a highway. At the time of the City Centre Master Plan preparation, a great part of the area situated north of Old Rauma and inside the determined World Heritage buffer zone was in an undefined and derelict state, since a precast-concrete factory that had been located in the area for decades had ceased operations. As the planning of the area progressed, the city of Rauma was of the opinion that almost any building or use of the area would mean an improvement, when compared to its current unclear status.

One key incentive for beginning the preparation of the City Centre Master Plan in the first place was to define the future use of this old industrial area, called Leikari-Lampola. The main issue to be dealt with was whether to reserve the area for combined residential and office use or for commercial use. The general principles supporting the latter option were decided on early in the planning process, even though not to the extent that was originally proposed by the main owner of the area, the Skanska real estate development company, which promoted the construction in the area of a conglomeration of three large hypermarkets. One downside of a project of this extent that was mentioned in the early reports analyzing the commercial use of the Leikari-Lampola area was the emptying of Old Rauma of businesses due to the likely concentration of smaller specialized shops, besides the hypermarkets, in the new commercial area north of Old Rauma.[87]

The City Centre Master Plan confirmed that no new hypermarkets would be built in Rauma. Instead it was felt that the two hypermarkets already based in Rauma should be allowed to expand their businesses in a large-scale shopping center development in the Leikari-Lampola area. The first of these, City Market, was already situated in the western part of the area, whereas the other, Prisma, was to relocate to the area from a few kilometers away. The Master Plan assigned 40,000 m² (gross area) of overall building rights to the area. The National Board of Antiquities was consulted about the plan, but in its comments the Board focused mainly on the delimitation of the World Heritage buffer zone.[88] Later on, the Board implied that it had been misled by the city of Rauma concerning the scale of the shopping center plans. The City Environmental Board, while commenting on the draft City Centre Master Plan, however, already at this stage, pointed out the potential controversy between the World Heritage buffer zone definition and the plans to bring the two large-scale commercial units into the very same area, thus potentially undermining the special status of Old Rauma as a World Heritage site.[89] In its response to the Environmental Board, the City Planning Office asserted that the building of the large-scale units so near the city center would support the continuing commercial use of Old Rauma. The planners also gave assurances that the selected solution would not mean a significant increase of floor area in commercial use in Rauma.[90]

The controversy that was to follow centered essentially around these two arguments. The first one was taken up by the National Board of Antiquities, the

Old Rauma Society, the Old Rauma shopkeepers' association and some individual entrepreneurs and property owners, whereas the second one was promoted, most importantly, by the city government with the support of the local newspaper. The real estate developer participated very little in public discussions. Thus, the older demarcations of dissonance concerning Old Rauma became ambiguous, as the Old Rauma shopkeepers joined forces with the conservationists with the common objective of protecting Old Rauma's liveliness from the potentially detrimental effect of the shopping center development. In the City Centre Master Plan phase, the groups with conservationist interests had been generally supportive of the planned commercial use of the Leikari-Lampola area when contrasted with the residential area option. This was founded on the notion that a large-scale residential use of the Leikari-Lampola area might lead to an unnecessary overemphasis on Old Rauma as a business area and consequently complicate maintaining its existing residential use and historical authenticity. Here we still find some of the traditional opposition to the commercial and residential uses in defining the liveliness of Old Rauma. In its remarks on the City Centre Master Plan, the Old Rauma Committee considered the size of the hypermarkets and the associated functions the most important issues for Old Rauma: "If the shopping centers will have small specialty shops in addition to the big stores, the competition can prove to be fatal for Old Rauma. On the other hand, well arranged connections from the market-area to Old Rauma can also bring more customers to the unique small shops of the area."[91] This view was shared by all those stakeholders who later came to oppose the proposed project – it was the scale and the realization of the project that came to be rejected as contrary to World Heritage value.

A detailed plan, adopted by the City Council in September 2004,[92] allowed 41,100 m^2 (gross area) of overall building rights for the developed area. Altogether 31,700 m2 were envisioned for the two large-scale units, 7,420 m^2 of which could be used for small, specialized shops. The rest of the 41,100 m^2 were reserved for the purposes of office use and specialized commerce requiring large amounts of floor space, such as that needed for furniture and car sales. In response to the critical feedback that had been received in consultations during the planning process, the final plan introduced regulations for the size and kind of small shops attached to the two hypermarkets, and a system for phasing with regard to the deployment of commercial space and its division for associated specialized stores.[93] The role of these restrictions, however, was quite marginal, since they were in force less than three years after the completion of the Prisma hypermarket in 2007.

The National Board of Antiquities now considered the shopping center development highly contradictory to Old Rauma's World Heritage status. During the preparation of the detailed plan, in contrast to the Master Plan drafting phase, the Board repeatedly showed reservations towards the project, and when the detailed plan was adopted by the City Council in September 2004 without any significant alterations, the Board decided to appeal to the Administrative Court, and later to the Supreme Administrative Court. The National Board of Antiquities urged the city to stand up for its international responsibilities based on the World Heritage Convention. It also saw the adopted detailed plan as being severely at odds with the recently adopted National Land Use Guidelines of 2000. The purpose of these guidelines, issued by the Finnish Government, is to promote nationally important planning objectives in regional and local planning. At the time of the shopping center planning, the objectives included, for example, sustainable development, a good living environment, the building of a coherent urban structure, and the promoting of land usage, which supports the protection of significant values of cultural heritage identified nationally. Furthermore, all land usage was to take into account the implementation of international commitments in heritage management.[94]

The shopping center debate in Rauma was actually the first time in Finland since the 1987 signing of the World Heritage Convention that the National Board of Antiquities publicly intervened in a local management and planning process in relation to a World Heritage inscribed site.[95] In various forums, the Board opposed the plan both on the grounds of townscape quality and for historical, structural, functional and social reasons. The building rights granted and the scale of the units were considered too extensive, large parking areas intrusive, and the overall aesthetic considerations of the project insufficient, even if the plan introduced some height limits and restrictions on advertisements. Most importantly, the Board expressed its concern that the shopping center development would change Old Rauma's structural and functional status in relation to the rest of the city, that is, its position as part of the daily commercial center of Rauma, turning it into either a sleeping residential suburb or a reserve for tourists. In both these cases the special characteristics of the area would be lost. The conclusion was that all this would significantly lessen Old Rauma's outstanding universal value and its associated liveliness.[96] Here we may find that the focus of considerations about liveliness in the context of Old Rauma was different in comparison to many other World Heritage cities trying to fight the trend of transformation from spaces of living into spaces for selling.[97]

During the planning process, similar arguments were also put forward by the Old Rauma Society, the Old Rauma shopkeepers' association and individual entrepreneurs. Even though all these different stakeholder groups had their own points of departure and interests in the process, one common feature of all the appeals was the expected devaluation of outstanding universal value from the point of view of the liveliness of Old Rauma.[98] As voiced by one shopkeeper interviewee, World Heritage was their "weapon to use" to resist the shopping center plans."[99]

Shopkeepers were the most visible group opposing the project at the local level. Their arguments were well summarized in an appeal to the Supreme Court written together with a prominent local architect, Jukka Koivula, who, since the 1970s, had been actively involved in planning for Old Rauma:

> This amount (17,700 m² of specialized commerce and services left in Old Rauma) is very critical in the respect that the area can still be considered as having liveliness in the terms of the UNESCO decision, if, for example, the market square can be revitalized to initiate the selling of groceries. Any diminishing would bring degeneration. Old Rauma was accepted onto the UNESCO World Heritage List in 1991 as a coherent urban entity, the

characteristics of which are wooden buildings of various ages with their yards, a street plan originating partially from medieval times and a lively urban community.[100]

The shopkeepers tried to establish, by reference to 'cold' numbers, the critical point of Old Rauma's commercial liveliness. Once again it was stated that the liveliness should not be understood only in terms of living but also in terms of commerce. Moreover, by arguing that "a special requirement is set by Old Rauma, the World Heritage status of which is not based only on visual attributes,"[101] shopkeepers questioned the understanding of outstanding universal value as a visual category only. For them outstanding universal value showed itself first and foremost as a functional and social value. In the minds of the shopkeepers the definition of outstanding universal value based on commercial liveliness had acquired a special role, and had become a constituent of the meaning of a place. Equally important, it was World Heritage that was now seen as a means to retain control of Old Rauma as an economic resource.

In its many pleas the city of Rauma justified the selected solution on community structural grounds. With the Leikari-Lampola plan the city of Rauma wanted to prevent the decentralization of hypermarkets, and the consequent decentralization of the urban structure, something that was also a key concern in the National Land Use Guidelines. This would support the commercial future and liveliness of Old Rauma as well.[102] In the city of Rauma's statements the project was considered in no way contradictory to the World Heritage buffer zone decision made only a year earlier. Or rather, in many of these statements the World Heritage status and value were left almost completely unstated. The belonging of the planned area to the World Heritage buffer zone was noted in the planning documents, but there was virtually no discussion of what this meant for the project or for Old Rauma. It is illustrative that the summary of the probable effects of the plan did not mention World Heritage or the buffer zone at all, even in the section labeled "Building protection."[103]

Furthermore, the liveliness of Old Rauma, even though stated as a concern elsewhere, was not discussed by the city government in connection with World Heritage. Instead, when mentioned, World Heritage was coupled with the question of the townscape effects of the project. What was at work here was, on the one hand, the downplaying of liveliness as an outstanding universal value in its own right, and a solely visual understanding of the concept on the other hand. The understanding of outstanding universal value as an aesthetic and visual category is discernible throughout the planning material, but most clearly so in the special building guidelines attached to the plan.[104] That World Heritage mostly was discussed in this particular section, and not in the others, also supports the conclusion that the primary role was given to the visual. In other words, what was implied was that by adhering to certain restrictions regarding the building and landscape (the height of construction, façade materials, fencing of the area, lighting and planting), the World Heritage value would remain intact. The building guidelines began with the City Centre Master Plan statement concerning the buffer zone as protection of the special status of Old Rauma in terms of cultural history and townscape. The rest of the guidelines, however, saw these conditions fulfilled only if the shopping center area was visually subordinate to Old Rauma, such that constructions were not too high to prevent views to or from Old Rauma. There was no elaboration on what the reference to cultural history meant. This understanding of World Heritage value was crystallized in the city of Rauma's plea to the Supreme Court:

> The area of Leikari-Lampola and Old Rauma are visually connected only at the church where the distance to the shopping center area is about 300 meters. Otherwise there is no visibility from Old Rauma of the shopping center. The central cityscape-related theme in the Leikari-Lampola plan is the respect for the church silhouette and the idea that the shopping center should be visually inferior to Old Rauma. The essentials in this case are the heights of the buildings, the borders and sides of the areas and the lighting, all of which are regulated in the town plan. [...] According to the city of Rauma the town plan amendment fulfills the requirements of the World Heritage buffer zone in terms of building rights, cityscape and scale of construction, and it does not diminish the special values of the built environment and respects the valuable attributes of the World Heritage site.[105]

The City of Rauma's position was supported by the Supreme Administrative Court, which in January 2006, after a 4–1 vote, ruled in favor of the shopping center construction (Figure 6.7). In its ruling, the Supreme Administrative Court stated that the townscape effects of the project were not considered detrimental to

Figure 6.7 The Prisma site with large parking areas in 2015.

Photograph by Hannu Vallas/Lentokuva.

Old Rauma's heritage values in a way which would prevent the implementation of the National Land Use Guidelines. This position was mostly justified by noting that the planned area was located 300 meters distant from Old Rauma, and that in the intermediate area high buildings already existed.[106] Here, it is of course possible to ask if the Supreme Court decision would have been different if the northern side of Old Rauma had been less effectively built up. It is also possible to question whether the Supreme Court's justification will hold true in a (hypothetical) situation somewhere in the future, when some of the building stock in between the two areas is demolished.[107]

The Supreme Court also took a stand concerning the presumable effects of the shopping center development on Old Rauma businesses. It clearly stated that the shopping center development would most likely have a negative effect on Old Rauma, and that the specific regulations as to the size and kind of small shops introduced in the plan could only delay, but in no way hinder, this line of development. It was also noted that regardless of the plan the current trends in commerce, at work in Rauma and elsewhere, were leading towards a formation of bigger units. Therefore, the Supreme Court concluded that bringing the two hypermarkets into the vicinity of Old Rauma could, on the one hand, lessen the prerequisites of Old Rauma businesses, but, on the other hand, help in maintaining the area as a functioning community. The Court pointed out that over the decades there had been altering uses for Old Rauma buildings depending on each current need, thus suggesting that commercial buildings could be again turned into residential buildings. Accordingly, the Court noted that it had not been shown that the shopping center development would endanger maintaining Old Rauma as a lively historic community with services and shops.[108]

The above discussion suggests four conclusions in particular. Firstly, it confirms the position of liveliness as a key meaning associated with Old Rauma. All stakeholders, including the city of Rauma, voiced their commitment to Old Rauma's liveliness; however, ways of maintaining the liveliness were again seen as having differing meanings. Secondly, on the basis of the above discussion, outstanding universal value shows itself to be an abstract and fluid concept, not easy to defend as part of local planning and land use debates. The Old Rauma example shows that the regulatory role of a World Heritage buffer zone in the case of actual planning in the vicinity of the World Heritage area may be minimal, even though it is also possible to point out several completely opposite examples, such as that of Vienna, as discussed in Chapter 2. One conclusion to be drawn from the Old Rauma example is that World Heritage was elaborated very little as a concept in its own right by some of the key stakeholders, such as the city government and the regional as well as the national Administrative Courts. The written definition of the buffer zone, even though included in a City Centre Master Plan, was too general a notion to become operational. The city of Rauma was able to state that during the planning process the special status of Old Rauma had been considered, and that the national authority responsible for World Heritage had been consulted. It seems that in Finland the forming of such a regulatory role would require the inclusion of World Heritage in national legislation. Among the States Parties to the World

Heritage Convention there exists no standard legislative or regulatory approach to ensure that the obligations to the Convention are met nationally. There are certain countries, such as Australia and the United Kingdom, that have introduced additional regulatory controls specifically for World Heritage. In the former, the World Heritage sites have been placed under special legal protection since the early 1980s, and in the latter the World Heritage sites are regulated under particular planning and policy guidance (*Policy Guidance from the Office of the Deputy Prime Minister*, 2005), according to which all significant development proposals affecting World Heritage areas require formal environmental assessment, "to ensure that their immediate impact and their implications for the longer term are fully evaluated."[109] In Finland, during the recent preparation of a national Building Protection Law, the National Board of Antiquities proposed that World Heritage sites should be specified as a group requiring special protection, but this proposal did not gather broader support nationally.[110]

Thirdly, there seems to be one particular meaning of outstanding universal value that is widely shared across the various levels of scale – namely that of outstanding universal value as a visual category, for which scale and style are the most important variables. Even though the World Heritage regime has expressed a commitment to broadening the range of values that it considers important in reference to urban heritage, the majority of urban threats discussed by the international stakeholders are seen as threats to visual integrity. When somewhat simplified, it may be claimed that as long as a particular development project is not visible to the World Heritage area, it does not pose a threat to the World Heritage value. This position was exemplified by the case of Vienna and may be discerned also in other urban contexts. In Rauma, the local stakeholders in favor of the shopping center building refused to consider anything other than construction visible to Old Rauma as an endangerment to outstanding universal value.[111]

The "lively community with various services, residential buildings and shops" was ultimately not regarded as part of outstanding universal value by the Courts. This suggests, fourthly, that the definitions of outstanding universal value based on functional, social or intangible aspects of a heritage place are much more difficult to make explicit and defend than the ones associated with more traditional heritage values. Equally difficult to transmit, it seems, is the idea that World Heritage can be something more than a phenomenon strictly limited within its borders.

Not writing to UNESCO: Local, national and global ownership of heritage

When considering the globalization of a place through a World Heritage designation three questions seem particularly important to answer. The first question is whether or not global concepts and issues become addressed in the local context after the designation. The second question is whether or not, following the designation, certain local matters become addressed, defined and even decided on at the international level, therefore leading to a partial transfer of power and agency away from the national and local levels.[112] The third, encompassing question relates to the relationship of the various heritage scales in the definition of World Heritage sites. This chapter has thus far mainly addressed the first question. It has been shown that World Heritage and outstanding universal value have become part of local negotiations over Old Rauma, and have been absorbed into the ongoing discourses on conservation, urban planning, commercial activities and identity formation, albeit in subtle and selective ways. While doing so, World Heritage has challenged, reshaped and further naturalized some representations of urban space and identity. There have also been direct consequences of the World Heritage listing for Old Rauma in terms of heritage practice: the most important example being the introduction of the concept of a buffer zone itself, which has no equivalent in the Finnish system of cultural heritage protection. The purpose of this section is to address the second and third questions: to discuss the shopping center debate in Rauma in terms of the coming of local issues to the global agenda, and in terms of the relationship of the different heritage scales. While continuing to focus on Old Rauma, I will occasionally make comparisons to two other World Heritage cities, Vienna and Dresden. Both these cities have experienced similar, although larger-scale, development versus conservation issues related to their World Heritage areas.

One rather surprising feature in the debate concerning the shopping center development in Rauma was that none of the stakeholders opposing the project, neither the National Board of Antiquities nor any of the local groups, at any stage of the events, contacted UNESCO formally, or even informally. This seems surprising since one of the main arguments used by these stakeholders was that the planned construction would mean a significant loss in terms of the outstanding universal value. It is also highly contradictory to situations in many other cities such as Dresden and Vienna, where the local NGOs and individuals contacted UNESCO immediately when controversies started to emerge in the planning process. The procedure of reactive monitoring in the World Heritage system is, in fact, largely dependent on the active role of locals in informing the World Heritage Committee and its advisory organizations concerning potential or actual threats affecting their localities inscribed as World Heritage.

It is also important to note that in the cases of Dresden and Vienna some of the local NGOs' objectives ended up as part of UNESCO's agenda, for example promoting the tunnel option as an alternative solution to bridge building in Dresden.[113] In Vienna the local conservationist NGOs were – in the end – more successful in their objectives than their counterparts in Dresden, but in both cases what was at work was a new form of local-global connection. In other words, the addition of global scale to heritage through World Heritage designation in these cities supported certain local claims for heritage over national and other local concerns. Carina Green has made similar observations in her study about the Swedish World Heritage mixed site Laponia where she sees Laponia as "an example of how the World Heritage status creates a platform of communication where the voice of the nation state is toned down on behalf of local and international voices," the local voice in that case being the voice and the agenda of Sami reindeer herders.[114] This re-established global-local connection would suggest that in certain cases the World Heritage system acts much in line with Roland Robertson's concept of "glocalism," a form of interconnected-ness of the global and local in the present era of globalization.[115] Furthermore, it would even suggest movement towards a reduced role for the nation states in the definition of heritage, in contrast to the argument in favor of the ever dominant role of the national in the definition of World Heritage.[116] For Old Rauma, however, this type of local–global connection failed to form.

A question that needs to be addressed, therefore, is why none of the stakeholders involved in the Rauma shopping center debate decided to establish a direct link with UNESCO. There are several possible explanations for this. Firstly, it may well be that very few if any of the stakeholders involved in the debate, with the exception of the National Board of Antiquities' representatives, even came to think of contacting UNESCO as a possible and valid option in furthering their own objectives. This may have been related to the limited knowledge about the World Heritage–related practices among locals. While some of my interviewees not professionally engaged in conservation interpreted World Heritage status as a means of external control,[117] the overall workings of the World Heritage system remained unclear for the major-ity of them. There was obscurity, for example, about the possible sanctions in the system, the practice of monitoring, the idea and role of the buffer zone, who can contact UNESCO, and what kind of threats the World Heritage Committee consid-ers worth discussing in the first place. Consequently, not writing to UNESCO may have been related to viewing the organization as an abstract stakeholder or, as one of the interviewees noted, as a "faceless collective" in the Old Rauma related pro-cesses: "It is so far away. It is hard to say whether there would be any sanctions."[118]

Secondly, what was also discernible in the debate was a reluctance to transfer authority away from the local level to the international, or even to the national level. In the discussions that I had with the local community, some interviewees showed an understanding with regard to World Heritage as a responsibility, a net-work and a support system, as eloquently put by one resident interviewee:

> [Through World Heritage] a network [is built] where people experience equality, in such a way that we belong to the same list. Then there is this respect, this kind of a global community, which reminds us that nature is a starting point for everything [...] And men have made some marks, monu-ments, places, nurtured some areas of cultures in such a way that they will remain for future generations. They should be taken care of. Like this little city of Rauma, since it is on the World Heritage List, these people get strength from it. It is concrete to be in a network and to be in interaction with unknown people. Be a part of a worldwide way of thinking and valuing and flow.[119]

In this statement, and in other similar ones, the interviewees expressed a strong commitment to Old Rauma. According to them, it was the common duty of people in Rauma to transmit Old Rauma to future generations of the world, although there were varying interpretations about what made up the most essential values to be passed on. To some the duty related particularly to the outstanding universal value stated in the World Heritage context, but for many it existed regardless of the new status. Some interviewees expressed the opinion that since the nomination of Old Rauma to the World Heritage List had been primarily a national-level decision, following the designation, the national government should have become more committed to Old Rauma's protection, especially in terms of government financial assistance. This was especially hoped for by the local conservation authorities, and should be seen as a continuation to the discussions in the 1980s, when there had been many (mainly unsuccessful) efforts to improve Old Rauma's funding nationally.[120] However, beyond these appeals for governmental financial assistance, what was striking was the strong emphasis on local responsibility. The shared duty to protect was not understood from the point of view of the international community being all together responsible for Old Rauma – or, resultantly, the international agent having a say in matters concerning Old Rauma. This position was clearly articulated in the text of the City Centre Master Plan, which stated that "Old Rauma has become not only a local but a worldwide heritage site, the responsibility for which, however, belongs to Finland, and to Rauma."[121] The question of whose duty it is to preserve is closely linked to the question of ownership and control. Whereas cultural heritage ideally should not be considered in terms of property rights and ownership at all,[122] the idea of ownership nevertheless has a central place in the conceptualization and use of heritage, and its resultant dissonance.[123] While emphasizing their duty alone to preserve, and by not approaching UNESCO at any stage, the local community articulated first and foremost their ownership of Old Rauma. This would suggest that the specificity of Old Rauma as a place has not significantly receded in the face of globalization's homogenizing forces, when viewed in terms of World Heritage.[124] The Old Rauma example shows that neither Old Rauma nor the values associated with it have become owned by the world community.[125]

The new local-global alliances may also raise concerns about their democratic nature, and may be regarded as a violation of local ownership and control over heritage. In Dresden the supporters of the bridge construction founded their central argument on the democratic legitimacy of the project. They placed the democratic deficit at the feet of UNESCO for not respecting the result of the local referendum supporting the bridge building, and with those local citizens' groups that actively worked to involve the World Heritage Committee.[126] In Rauma, too, even though more implicitly, some perceived UNESCO as an undemocratic stakeholder.[127] Still, while certain democratic flaws may be identified in relation to World Heritage, as asserted by Natasha Affolder, "the Convention also embodies certain democratic strengths when compared with alternative regimes for global governance," such as industry-led initiatives.[128]

There was also unwillingness to hand Old Rauma over to international consideration on the part of the national conservation authorities, despite all the references made to the World Heritage Convention. During the planning and legal processes the National Board of Antiquities indicated that the implementation of the shopping center plans to their full extent might result in a reassessment of Old Rauma's World Heritage status, but the Board did nothing concrete to bring the debate to the World Heritage Committee's attention. Again, we may find many opposite examples of a similar situation in other countries. For example, in the situation of a conflict between the city and the state authorities on the construction of a skyscraper within the buffer zone of the World Heritage site of the Historic Centre of Riga, the request for international monitoring was carried out by the State Inspection for Heritage Protection in Latvia.[129] In Finland, the National Board of Antiquities wanted to limit the buffer zone debate strictly to the national level. This was the case even though amidst the active phase of the Court appeals, the Board had to report to the World Heritage Committee, within the framework of the regional periodic monitoring exercise, about the implementation of the Convention in the context of Finnish World Heritage sites, including Old Rauma. In the brief report, the Board identified environmental and development pressures as main factors affecting Old Rauma's state of conservation. Furthermore, it acknowledged that as a living city center Old Rauma would always be "an area for many differing needs and ideas," and that the commercial prerequisites in the city were changing, which would also have influences on the city center, and "maybe not only positive" ones. The buffer zone, however, was described as an operational tool, able to prevent potential threats: "The idea is that all the activities, new buildings and traffic should be seen from the point of view of Old Rauma and its World Heritage values. The buffer zone should protect the surroundings of the old town from too high or too big buildings and disturbing activities."[130] The then topical project of shopping center building was not specified in any way in the report.[131] When later interviewed, the World Heritage coordinator in Finland, Margaretha Ehrström, noted that "there should be very substantial reasons, when going to the global arena; things can easily get out of proportion."[132] This suggests that even the National Board of Antiquities ultimately did not think that the shopping center building would result in an irreversible loss of outstanding universal value. It also indicates the Board's uncertainty about the sensitivity of the World Heritage system to local and national characteristics. The European Union's Natura program perhaps acted as one negative precedent. In Finland, in the late 1990s, this program, dealing with nature protection, had resulted in a major environmental conflict between local and national stakeholders.[133]

Today, many World Heritage cities are closely followed, not just from within but increasingly from outside national borders. A case in one country is quickly contrasted with another case from another country. The role of the Internet is essential in evoking this international attention. The developments taking place in Vienna and Dresden were reported widely. In Dresden, both the supporters and opponents of the bridge construction maintained extensive web pages, both in German and English, to inform people about their agenda. Collective memory today operates in an increasingly transnational context.[134] Undoubtedly, if such transnational spaces of memory and heritage truly exist,[135] many World Heritage cities make good examples of them.

Still, interestingly enough, there are certain cities, such as Rauma, which seem to avoid most of the international gaze, even when part of World Heritage. That UNESCO was not informed about the major construction in the World Heritage buffer zone in Rauma is one thing, but another is that the organization did not hear about the matter from any other source. In part, this is a hierarchical issue of size and former fame. When compared to Dresden and Vienna, Rauma obviously operates in a different league. Dresden and Vienna are cities that have world fame with or without World Heritage status: as major cities in their region, as important destinations for international tourism, and as internationally well-known examples of conservation. When it comes to smaller and internationally less-well-known cities, the World Heritage status seems to become ever more important in terms of urban regeneration, outward acknowledgement, generating additional funding for conservation and boosting local identity. At a time of dissonance, however, it is often the big, well-known cities that are brought to the forefront as representatives of the heritage of mankind and its neglect. As asserted by Marc Askew, many "controversial internal issues and struggles do not enter the purview of UNESCO's deliberating and policy-making bodies," which in turn reinforces the assumption "that World Heritage sites are the uncontested icons of each country's individual 'culture' and 'tradition.'"[136] Whatever the reason that UNESCO was not informed about the major building project in Rauma – because it was major for Old Rauma, even though different in scale than building a skyscraper in Vienna or constructing a car bridge over the Elbe in Dresden – from UNESCO's perspective it seems somewhat alarming that there are significant projects going on in World Heritage–designated sites that it does not know anything about. It is difficult to say whether or not and how the World Heritage Committee might have decided to get involved in the debate in Rauma. Nevertheless, the Rauma example does show the enormous task that the World Heritage organization is facing at the moment as it tries to manage over 1,000 (with more to come) World Heritage sites.[137]

In certain respects the Old Rauma case study supports the notion of the enduring centrality of nation states in the World Heritage framework. World Heritage clearly acted as an empowering agenda for the national conservation agency to oppose the local development project. Still, it ought to be noted that the national ownership of heritage in this case meant primarily that of the national courts. It seems that whatever may be the national system for protection of World Heritage sites, when controversies arise, it is ultimately the national courts that decide on how the outstanding universal value should be interpreted. In its ruling, the Supreme Administrative Court of Finland widely referred to the objectives outlined in the 2000 National Land Use Guidelines. When justifying its decision, the Court, however, did not address the Guidelines' objective which stated that all land use should take into account the international commitments in the field of heritage management.[138] Therefore one conclusion must be that in Finland, as in many other countries, very limited additional protection has actually been afforded to heritage sites following their World Heritage designation. With its decision the Court favored the legality of a local planning process over international commitments.

Even though having overly categorical divisions between global, national and local in the framework of World Heritage is somewhat unfruitful, since the claims at various levels are simultaneously true,[139] it nevertheless seems pertinent to end this chapter by returning to the question to whom does World Heritage belong. According to UNESCO's construction, it belongs to everyone – in addition to being a shared global responsibility, World Heritage reveals itself as a shared right and interest of humankind regarding a common resource. These visions seem to be gaining more ground within the framework of the global memorial culture as UNESCO moves to strengthen its position as an international stakeholder. That many World Heritage sites continue to be managed under national legislation and regulatory systems does not automatically mean that UNESCO and the World Heritage regime are powerless stakeholders. They have moral suasion, and the power to determine the framework for heritage discussions. However, in light of the Old Rauma example, it seems sufficient to say that World Heritage as a factor of globalization did not mark any global ownership of heritage. The agency for Old Rauma remained firmly at the local and national levels. The outstanding universal value may be treated in a uniform way by UNESCO and ICOMOS, but the local communities' and the nation states' hold on their heritage remains strong. Depending on the viewpoint, this may be considered a success or a failure from UNESCO's perspective. Local people's participation in the management of World Heritage sites is currently something that is strongly encouraged by the World Heritage Committee and ICOMOS. What they have to accept is that local community participation does not always have positive outcomes in light of their own objectives.

Notes

1   Heino, Ulla, *Rauma – Idylliä ja tehokkuutta 1875–2000* (Rauma: Rauman kaupunki, 2002), 277–304.
2   For discussion of the consequences of the World Heritage status for individual cities see, for example, Seppänen, *Global Scale*; Michael Hitchcock, "Zanzibar Stone Town joins the imagined community of World Heritage sites," *International Journal of Heritage Studies* 8: 2 (2002): 153–166; Melanie Smith, "A critical evaluation of the global accolade: the significance of World Heritage site status for Maritime Greenwich," *International Journal of Heritage Studies* 8: 2 (2002): 137–152; Evans, "Living;" Tanja Vahtikari, "Urban Interpretations of World Heritage: Re-defining the City," in *Reclaiming the City: Innovation, Culture, Experience*, ed. Marjaana Niemi and Ville Vuolanto (Helsinki: SKS, 2003), 63–79; Ute Klimpke and H. Detlef Kammeier, "Quedlinburg – 10 years on the World Heritage List: east-west transformations of a small historic town in central Germany," *International Journal of Heritage Studies*, 12: 2 (2006): 139–158; Klusáková, "Between urban."
3   Michael C. Hall, "Implementing the World Heritage Convention: what happens after listing?" in *Managing World Heritage sites,* ed. Anna Leask and Alan Fyall (Amsterdam: Elsevier, 2006), 21.
4   See also Nilsson Dahlström, *Negotiating Wilderness*, 21. In my analysis I have focused primarily on arguments officially presented by the various groups, therefore according less room to the variations that exist at the intra-group level.
5   Michael James Miller, *The Representation of Place: Urban Planning and Protest in France and Great Britain, 1950–1980* (Aldershot: Ashgate, 2003), 29.
6   Larkham, *Conservation and the City*; Ashworth, "The conserved European city," 261–286; Ashworth and Tunbridge, *The Tourist-historic City*; Borsay, *The Image of Georgian Bath*; Jordan, *Structures of Memory*; Hagen, *Preservation, Tourism and Nationalism.*
7   For early stages of heritage-making for Old Rauma see, Tanja Vahtikari, "Merkityksin rakennettu Vanha Rauma: suomalaisen historiallisen kaupungin varhainen määrittely valintoina ja vuoropuheluna 1900–1970," in *Muistin kaupunki. Tulkintoja kaupungista*

muistin ja muistamisen paikkana, ed. Katri Lento and Pia Olsson (Helsinki: Finnish Literature Society, 2013).

8 See, for example, *Vanhan Rauman kaavarunko* 31.12.1972. Vanhan Rauman asemakaavatoimikunnan 1969–1972 asiakirjat. Erinäisten kaavoitukseen ja yleisiin töihin liittyvien toimikuntien ja työryhmien asiakirjat 1969–1975. Cff: 4. KV/KH/ KK. Rauma City Archives (RCA).

9 Pöytäkirja ns. vanhan eli I kaupunginosan saneerauksesta Rauman kaupungin edustajien ja mainitun kaupunginosan talon- ja tontinomistajien välillä 22.3.1961. Vanhan kaupungin saneeraustoimikunta 1961. Erinäisten kaavoitukseen ja yleisiin töihin liittyvien toimikuntien asiakirjat 1959–1970. Cff: 3. KV/KH/KK. RCA.

10 Seija Jumppanen, *Die Innere Differenzierung der Stadt Rauma* (Turku: Turun yliopisto, 1973).

11 Vahtikari, "Miten Vanhasta Raumasta," 103; Interview, Old Rauma resident, 25 October 2002.

12 For a similar account see Borsay, *The Image of Georgian Bath,* 168–184.

13 Rauman keskustan asemakaavakilpailun palkintolautakunnan pöytäkirja, 29.1.1965. Keskustan asemakaavakilpailutoimikunta 1963–1965. Erinäisten kaavoitukseen ja yleisiin töihin liittyvien toimikuntien asiakirjat 1959–1970. Cff: 3. KV/KH/KK. RCA.

14 Jokilehto, *A History,* 303.

15 Raphael, *Theatres of Memory,* 160–161; Jordan, *Structures of Memory,* 42. Many of the public protests in favor of cultural heritage in the 1970s and 1980s ended up being failures, even though the actual buildings were preserved. This was due to gentrification of the protected areas – the original residents of these areas, promoting protection, often could not afford to live in them any longer after the renovation. See, for example, Smith, *Uses of Heritage,* 24–25. This was not however the case with Old Rauma where this form of gentrification did not take place.

16 See, for example, Larkham, *Conservation and the City,* 39, 42.

17 *Vanhan Rauman kaavarunko* 31.12.1972. RCA.

18 Interview, former local planning authority, 27 March 2002; Interview, national conservation authority, 21 February 2002.

19 Rakennuskulttuurin luokittelun tarkistus. Museovirasto, rakennushistorian osasto. *Vanha Rauma, asemakaavan muutosehdotus ja rakentamista ohjaava kokonaissuunnitelma.* Rakennusviraston kaavoitusosasto, asemakaava-arkkitehti Reino Joukamo ja toimistoarkkitehti Jukka Koivula, 1.4.1980. Kartta 13. Rakennus- ja kiinteistölautakunnan / teknisen lautakunnan pöytäkirjat liitteineen 1974–1981. Tekninen lautakunta / tekninen virasto (TEKLA). RCA.

20 *Länsi-Suomi* 12 April 1973.

21 *Länsi-Suomi* 6 March 1973 and 7 March 1973.

22 See, for example, *Länsi-Suomi* 3 May 1974 and 3 October 1974.

23 *Länsi-Suomi* 28 March 1975.

24 *Vanha Rauma, asemakaavan muutosehdotus ja rakentamista ohjaava kokonaissuunnitelma,* 1.4.1980. TEKLA. RCA; Koivula, Jukka, et al., *Vanha Rauma – Old Rauma* (Rauma: Rauman museo, 1992).

25 Vahtikari, "Miten Vanhasta Raumasta," 108.

26 A. B. Tammivaara, "Rauma – piirteitä sen kehityksestä ja uusista suunnitteluista," in *Rauma. Retkeilijä.* No. 5 (Helsinki: Finnish Literature Society, 1938), 76.

27 Rauman keskustan asemakaavakilpailun palkintolautakunnan pöytäkirja, 29.1.1965. KV/KH/KK. RCA. See also Keskustan asemakaavakilpailun aineisto, palkitut ja lunastetut ehdotukset 1964. Ha: 1. Hallinto-osasto (HO). RCA.

28 *Vanha Rauma, asemakaavan muutosehdotus ja rakentamista ohjaava kokonaissuunnitelma,* 1.4.1980. TEKLA. RCA; Koivula et al., *Vanha Rauma – Old Rauma.* For the pedestrian street debate see, for example, *Länsi-Suomi* 24 March 1978, 29 March 1978, 30 March 1978, 31 March 1978, 9 April 1978, 12 April 1978 and 14 April 1978.

178   *Outstanding universal value*

29 See, for example, Rauman kaupunki, tekninen lautakunta 1.11.1978, 1715 § & Liite A. Vanhan Rauman tontinomistajien ja liikkeenharjoittajien yhteiskirjelmä Rauman Rakennus- ja kiinteistölautakunnalle. 30.1.1978. TEKLA. RCA; Kaupunginvaltuusto 29.3.1982, No 20, Vanhan Rauman liikenteestä ja muusta johtuvien häiriöiden vähentäminen, *Rauman kaupungin kunnalliskertomus* 1982, 263–283.
30 Interview, Old Rauma property owner and former shopkeeper, 7 November 2002; Interview, Old Rauma shopkeeper, 7 November 2002.
31 *Länsi-Suomi*, 24 March 1978 and 29 March 1978.
32 *Länsi-Suomi*, 28 May 1981.
33 See, for example, Vanhan Rauman Erityiselin 5/82 6.8.1982. Kokouspöytäkirjat 1982–2003. Old Rauma Centre (OCR).
34 *Länsi-Suomi*, 17 April 1985.
35 *Vanha Rauma, asemakaavan muutosehdotus ja rakentamista ohjaava kokonaissuun-nitelma*, 1.4.1980. TEKLA. RCA; Koivula et al., *Vanha Rauma – Old Rauma*.
36 Most of these buildings were never demolished, though, but were successfully "integrated into their surroundings" with their functions changed. Koivula et al., *Vanha Rauma – Old Rauma*, 48.
37 For opposite early stage examples of the cities of Lübeck and Lima, see Vahtikari, "Urban interpretations," 65–66; Seppänen, *Global Scale*, 45.
38 Vahtikari, "Urban interpretations."
39 ICOMOS, Meeting for the harmonization of tentative lists of cultural proper-ties of Northern European countries, Bergen (Norway), May 24, 1986, Appendix: Typological and chronological framework proposed by Sweden. Hbb1. 3, UNESCO, Maailmanperintöasiat 1972–1997, National Board of Antiquities of Finland (NBA), Helsinki; Vahtikari, "Miten Vanhasta Raumasta," 108. For the Nordic harmonization of national tentative lists, see Vahtikari, "From National to World Heritage via the Regional" (forthcoming).
40 Nomination to World Heritage List, Old Rauma, Finland, 13.9.1990. World Heritage Centre, World Heritage Sites, Nomination files (1978–1997), Europe vol. I (CD in author's possession).
41 Ibid.
42 UNESCO, World Heritage Committee, Fifteenth Session (Carthage, 9–13 December 1991), Report of the Rapporteur, SC-91/CONF.002/15, 12 December 1991, p. 27–28, http://whc.unesco.org/archive/1991/sc-91-conf002-15e.pdf.
43 ICOMOS evaluation for the nomination of the World Heritage property, "Rauma," Finland, No 582, May 1991.
44 A retrospective Statement of Significance for Old Rauma was accepted in the 2014 session of the World Heritage Committee. UNESCO, Thirty-Eighth Session, Doha, Qatar (15–25 June 2014), 8E: Adoption of retrospective Statements of Outstanding Universal Value, Paris, 30 April 2014, WHC-14/38.COM/8E, p. 60, http://whc.une-sco.org/archive/2014/whc14-38com-8E-en.pdf.
45 ICOMOS evaluation for the nomination of the World Heritage property, "Rauma," Finland, No 582, May 1991.
46 There was in fact not a single local interviewee among the total of 22 who did not integrate the concept of liveliness into her/his definition of Old Rauma and its out-standing universal value.
47 See for example, Rauman kaupunki, kaavoitus, *Rauman yleiskaava* (Rauma Master Plan) 2.7.2003, Annex 1; Report on the buffer zone of Old Rauma World Heritage site. National Board of Antiquities, Finland 23.1.2009. In the possession of Margaretha Ehrström.
48 For similar discussion, see Askew, "The Magic List," 38–39.
49 Anna Nurmi-Nielsen, "Vanha Rauma – kulttuuriperintö meidän käsissämme," *Länsi-Suomi*, 12 March 1995.

50 *Länsi-Suomi*, 29 January 1995. See also *Länsi-Suomi*, 28 June,1994; *Uusi Rauma*, 27 April 1994.

51 Kauppakamarin ja Vanhan Rauman yrittäjien järjestämä keskustelutilaisuus Vanhan Rauman kehittämisestä ja kaupan näköaloista Raumalla. Muistio 29.1.1991. Neuvoa-antavien kokousten ja vastaavien pöytäkirjat ja muistiot liitteineen 1989–1991. CC4. KV/KH/KK. RCA.

52 See also Adams, "The politics of heritage."

53 Interview, mayor of Rauma, Pentti Koivu, 9 April 2002. For example, in 2000 the city of Rauma launched a city center redevelopment and enhancement project in which Old Rauma, and improving its commercial liveliness, were granted a prominent role.

54 Interview, Old Rauma real estate owner and former shopkeeper, 7 November 2002; Interview, Old Rauma shopkeeper, 3 November 2003.

55 Interview, Old Rauma shopkeeper, 6 November 2003.

56 Elisa El Harouny, ed., *Suomalaisia puukaupunkeja. Hoito, kaavoitus ja suojelu* (Helsinki: Ympäristöministeriö, 1995), 22, 24; Pekka Kärki in *Länsi-Suomi*, 18 March 1995.

57 Interview, local conservation authority, 24 November 2003.

58 Interview, local conservation authority, 5 November 2003.

59 Laura Tuominen, "'Museoiminen' metaforana ja rakennussuojelukriittisenä argumenttina," in *Rakkaudesta kaupunkiin*, ed. Renja Suominen-Kokkonen (Helsinki: Taidehistorian seura, 2004), 94.

60 El Harouny, *Historiallinen puukaupunki*, 233–234.

61 Interview, local conservation authority, 24 November 2003.

62 Interview, Old Rauma resident and Old Rauma Society activist, 27 November 2002; Interview, Old Rauma shopkeeper, 28 November 2002.

63 Interview, municipal politician, 25 October 2002.

64 For discussion see, Tuomi-Nikula, Outi, "Alasaksalainen hallitalo kertoo pientä historiaa. Talo eurooppalaisen etnologian tutkimuskohteena Saksassa," in *Historioita ja historiallisia keskusteluja*, ed. Sami Louekari and Anna Sivula (Turku: Turun historiallinen yhdistys, 2004), 265.

65 Interview, Old Rauma real estate owner and former shopkeeper, 7 November 2002.

66 See, for example, Vanhan Rauman asemakaavan toteuttamista ohjaava erityiselin (Old Rauma Committee) (VREe) 1/92 15.1.1992, 1/94 23.3.1994, 3/94 23.5.1994, 3/95 15.9.1995. Kokouspöytäkirjat 1982–2003. Old Rauma Renovation Centre (OR Centre), Rauma.

67 In this development the role of the Old Rauma Renovation centre, established in 1995, and the role of an advisory architect working in the area were decisive.

68 VREe 7/99, 25.08.1999. Kokouspöytäkirjat 1982–2003. OR Centre.

69 Interview, local conservation authority, 5 November 2003.

70 See, for example, *Uusi Aika*, 2 August 1983.

71 *Länsi-Suomi*, 5 October 1994, 26 April 1995 and 1 December 1995.

72 Interview, municipal politician, 25 October 2002; Interview, Old Rauma shopkeeper, 6 November 2003.

73 Minna Linnala and Liisa Nummelin, *Toiveet ja todellisuus. Satakunnan rakennusperinnön hoito-projekti 1998–2000* (Pori: Satakunnan Museo, 2000).

74 Ibid., 80, 82.

75 Interview, local conservation authority, 5 November 2003.

76 For discussion in the Old Rauma Committee see VREe 3/98, 20.08 1998 & 2/99, 4.2.1999. Kokouspöytäkirjat 1982–2003. OR Centre.

77 It has been noted, for example, that the medieval city of Visby was exposed to a large-scale restoration before its World Heritage nomination in order to make it look more medieval. Ronström, Owe, "Kulturarvspolitik. Vad skyltar kan berätta," in *Kritisk etnologi. Artiklar till Åke Daun*, ed. Barbro Blehr (Stockholm: Prisma, 2000), 64.

78 For discussion see Ashworth and Tunbridge, *The Tourist-historic City*; Anja Kervanto Nevanlinna, "Classified urban spaces: who owns history of Helsinki South Harbour?," in *Identities in Space. Contested Terrains in the Western City since* 1850, ed. S. Gunn and R. J. Morris (London: Ashgate, 2001), 19–38.

79 See, especially, Interview, Old Rauma resident and a member of the Old Rauma Society, 7 November 2002.

80 Klusáková, "Between urban," 207–208.

81 Interview, local conservation authority, 5 November 2003.

82 Rauman kaupunki, kaavoitus, *Rauman keskustan osayleiskaava* (City Centre Master Plan). Osayleiskaavan selostus, joka koskee 16.1.2003 päivättyä osayleiskaavakarttaa. Because of appeals, the City Centre Master Plan did not gain a legal status. These appeals, however, did not concern the buffer zone delimitation.

83 Ibid., 140.

84 UNESCO, World Heritage Committee, Thirty-third Session (Seville, Spain 22–30 June 2009), Report of the decisions, WHC-09/33.COM/20, Seville, 20 July 2009, p. 222–223, http://whc.unesco.org/archive/2009/whc09-33com-20e.pdf; UNESCO, World Heritage Committee, Thirty-third Session (Seville, Spain 22–30 June 2009), Evaluations of cultural properties, WHC-09/33.COM/INF.8B1.Add, p. 52–53, http://whc.unesco.org/archive/2009/whc09-33com-inf8B1ADDe.pdf.

85 Interview, the National Board of Antiquities' World Heritage coordinator Margaretha Ehrström, 23 November 2010.

86 Another area with a significantly different type of profile within the buffer zone area is the western side of Old Rauma, which is mainly a business and administrative area. The area, stretching to the bus station area towards the northern part of the buffer zone, is currently under re-planning, also highly controversial.

87 Rauman kaupunki, kaavoitus, *Leikari-Lampolan asemakaavan muutos* (AK 01-084), 8.9.2004, 13; Rauman kaupunki, kaavoitus, *Rauman keskustan osayleiskaava*, 16.1.2003, 53–56.

88 Museoviraston lausunto 23.8.2002: Rauman osayleiskaavaluonnos; Museoviraston lausunto 17.6.2003: Alustava Rauman yleiskaava-ehdotus ja alustava Rauman keskustan osayleiskaavaluonnos. Rauma IV: 1996–Topografinen arkisto. NBA.

89 Rauman kaupunki, kaavoitus, *Rauman keskustan osayleiskaava*, 16.1.2003, 117.

90 Ibid., 109, 119.

91 VREe 2/2002, 25.3.2002, Lausunto Rauman yleiskaavasta ja keskustan osayleiskaavasta. Kokouspöytäkirjat 1982–2003. OR Centre.

92 Rauman kaupunginvaltuusto 27.9.2004, 104§. KAN: 433/2004.

93 *Leikari-Lampolan asemakaavan muutos*, 8.9.2004, 41.

94 Valtioneuvosto, Valtioneuvoston päätös valtakunnallisista alueidenkäyttötavoitteista, November 30, 2000, Helsinki, accessed May 7, 2012, www.ymparisto.fi/download. asp?contentid=94382&lan=fi.

95 For nationwide publicity, see, for example, *Helsingin Sanomat*, 25 October 2004. Years earlier, the Board had successfully defended – by referring to the World Heritage status – the surroundings of the Fortress of Finland from the expansion of a waterway by mining one of the nearby islands. This, however, did not cause a public debate. Interview, national conservation authority, 29 January 2002.

96 "Museovirasto Korkeimmalle hallinto-oikeudelle 3.5.2005, Turun hallinto-oikeuden 4.4.2005 päiväämä päätös nro 05/0102/1 asemakaavan valitusasiassa," accessed 1.9.2012, www.nba.fi/fi/leikarilampola_valitus; Museoviraston lausunto 29.4.2004: Radanvarsitien asemakaavan muutos, Rauma. Rauma IV: 1996–Topografinen arkisto. NBA; Museoviraston lausunto 26.8.2004: Leikari-Lampolan asemakaavan muutosehdotus, Rauma. Rauma IV: 1996– Topografinen arkisto. NBA.

97 See, for example, Evans, "Living," 119.

98 The Old Rauma Society in its own appeal argued that the city government should bear its responsibility for "our World Heritage site remaining a World Heritage site.

It has been granted this status precisely because it has stayed a lively entity with its inhabitants and businesses." Vanha Rauma Yhdistys Rauman kaupunginhallituksen kaavoitusjaostolle 11.7.2004: Muistutus koskien Leikari-Lampolan asemakaavan muutosehdotusta. *Leikari-Lampolan asemakaavan muutos*, 8.9.2004, Liite 12. See also KHO: 2006:2, 26.1.2006, 1236/1/05, accessed April 21, 2012, www.kho.fi/paatokset/34553.htm.

  99 Interview, Old Rauma shopkeeper, 28 November 2002.
100 KHO: 2006:2, 26.1.2006, 1236/1/05, accessed April 21, 2012, www.kho.fi/paatokset/34553.htm.
101 Ibid. See also Vanhan Rauman liikekeskusyhdistys Ry 12.7.2004: Muistutus Leikari-Lampolan asemakaavaehdotuksesta. Laatineet Jukka Koivula ja Erkki Railio. *Leikari-Lampolan asemakaavan muutos*, 8.9.2004, Liite 12.
102 Interview, planning authority, 6 November 2003.
103 *Leikari-Lampolan asemakaavan muutos*, 8.9.2004, 44.
104 Ibid., Liite 11.
105 Kaupunginhallituksen kaavoitusjaoston kokous 2.6.2005, 93§: Vastine Museoviraston Leikari-Lampolan asemakaavan muutoksesta (Ak 01-084) Korkeimmalle hallinto-oikeudelle tekemään valitukseen, accessed August 31 August, 2012, www.rauma.fi/ktweb/.
106 KHO: 2006:2, 26.1.2006, 1236/1/05, accessed April 21, 2012, www.kho.fi/paatokset/34553.htm.
107 Lauri Jääskeläinen, "KHO:n ja hallinto-oikeuksien päätöksiä," *Rakennettu Ympäristö* 2 (2006), 70, accessed January 12, 2016, www.rakennustieto.fi/lehdet/ry/index/lehti/P_249.html.
108 KHO: 2006:2, 26.1.2006, 1236/1/05, accessed April 21, 2012, www.kho.fi/paatokset/34553.htm.
109 Hall, "Implementing," 24–25.
110 Interview, national conservation authority, 23 November 2010.
111 *Länsi-Suomi* editorial, 27 January 2006.
112 Dahlström, *Negotiating Wilderness*, 266.
113 UNESCO, World Heritage Committee, Thirtieth Session (Vilnius, Lithuania, 8–16 July 2006), Item 7 of the Provisional Agenda: Examination of the State of Conservation of World Heritage properties. WHC-06/30.COM/7B, Paris, 9 June 2006, p. 197–199, http://whc.unesco.org/archive/2006/whc06-30com-7bE.pdf; Birgitta Ringbeck, "Die Entscheidung zum Dresdener Elbtal. Ein Kurzbericht zur 33. Sitzung des UNESCO-Welterbekomitees vom 22. bis 30. Juni 2009 Sevilla, Spanien," accessed December 15, 2012, www.unesco.de/3652.html.
114 Green, *Managing Laponia*, 90. For a similar discussion see Seppänen, *Global Scale*, 132, who states that the designation of the Historic Centre of Lima as World Heritage site diminished the sovereignty of the nation state in favour of power on other scales.
115 Roland Robertson, "Glogalization: Time-Space and Homogeneity-Heterogeneity," in *Global Modernities*, ed. M. Featherstone, S. Lash and R. Robertson (London: Sage, 1995), 25–44; Green, *Managing Laponia*, 89–90.
116 See, for example, Ashworth and van der Aa, "Strategy and Policy;" Askew, "The Magic List."
117 Interview, Old Rauma resident and Old Rauma Society activist, 27 November 2002. See also Interview, Old Rauma resident 26 November 2002.
118 Interview, Old Rauma shopkeeper, 6 November 2003.
119 Interview, Old Rauma resident and Old Rauma Society activist, 27 November 2002.
120 Koivula, Jukka, "Vanha Rauma – Suojelukaavan jälkihoito," in *Rakennussuojelu*, ed. Jussi Rautsi and Kaija Santaholma, 128–129. Helsinki: Ympäristöministeriön Kaavoitus- ja rakennusosasto, 1987. See also *Länsi-Suomi*, 5 February 1992.
121 Rauman kaupunki, kaavoitus, *Rauman keskustan osayleiskaava*, 16.1.2003, 16.

122 K. J. Warren, "A philosophical perspective on the ethics and resolution of cultural property issues," in *The Ethics of Collecting Cultural Property: Whose Culture?*, ed. P. M. Messenger (Albuquerque: University of New Mexico, 1999), 1–25.

123 Lowenthal, *The Heritage Crusade*, 226.

124 For a common view in contemporary anthropology which opposes the one-sided fears of globalization as cultural homogeneity see, for example, Tsing, "The Global situation." See also Green, *Managing Laponia*, 88.

125 Cf. Carman, *Against Cultural Property*.

126 Bürgerinitiative Pro Waldschlößchenbrücke, accessed 15 January 2015, www.neue-waldschloesschenbruecke.de/unesco/unesco.htm.

127 See, for example, Interview, Old Rauma shopkeeper, 6 November 2003.

128 Affolder, "Democratizing."

129 UNESCO, World Heritage Committee, Thirtieth Session (Vilnius, Lithuania 8–16 July 2006), Item 11 of the Provisional Agenda: Periodic Reports. Sub-regional Periodic Reports for Europe, WHC-06/30COM/INF.11A, Paris, 23 June 2006, p. 139, http://whc.unesco.org/archive/2006/whc06-30com-inf11Ae.pdf.

130 Raportti Suomen maailmanperintökohteiden tilasta. Yhteenveto UNESCO:n kyse-lykaavakkeesta. Museovirasto Opetusminsteriölle 10.10.2005; Liite: State of con-servation of Old Rauma. Rauma IV: 1996– Topografinen arkisto. NBA. See also UNESCO, State of Conservation of World Heritage Properties in Europe, Section II, Finland, Old Rauma, accessed February 24, 2012, http://whc.unesco.org/archive/periodicreporting/EUR/cycle01/section2/582-summary.pdf.

131 Ibid.

132 Interview, the National Board of Antiguities' World Heritage coordinator Margaretha Ehrström, 23 November 2010.

133 Annukka Oksanen, *Paikallisuuden ja kansainvälisyyden kohtaaminen luonnonsuo-jelussa. Tapaustutkimuksena Natura 2000 -ympäristökonflikti Lounais-Suomessa* (Turku: Turun yliopisto, 2003).

134 Jordan, *Structures of Memory*, 193; Aline Sierp, *History, Memory, and Trans-European Identity. Unifying Divisions* (New York: Routledge, 2014).

135 Janina Fuge, "The local, the national – and the transnational? Spatial dimensions in Hamburg's memory of World War I during the Weimar Republic," in *Transnationalism and the German City*, ed. Jeffry M. Diefendorf and Janet Ward (New York: Palgrave Macmillan, 2014), 182–183.

136 Askew, "The Magic List," 38.

137 With regard the World Heritage List 'inflation' as regards the lack of management of already inscribed sites, see Ashworth and van der Aa, "Strategy and Policy," 155.

138 KHO: 2006:2, 26.1.2006, 1236/1/05, accessed 21.4.2012, www.kho.fi/paatokset/34553.htm.

139 See also Seppänen, *Global Scale*, 133.

7 Conclusions

> Until recently, few, if any, of the properties which make up the cultural heritage could be described as having significance outside their own immediate sphere of culture or influence. Indeed, some of the most remarkable products of human ingenuity and faith have had to be "rediscovered" and rescued from the encroaching forest after they had been abandoned and forgotten by the descendants of their creators. The whole concept of a world heritage is relatively new and depends upon an increasing awareness of the shared burdens and responsibilities of mankind as well as upon modern methods of transport and communication. It seems right, therefore, that the World Heritage Committee should avoid restricting its choices to the best known properties, but should also include these other properties, perhaps little known, but with great potential for aesthetic, educational and scientific value if made known to a wide public.[1]

The above statement, made by ICOMOS at the early stage of the implementation of the World Heritage Convention, summarizes well some of the main ideas linked to the concept of World Heritage over the years. World Heritage has been, and continues to be, about the themes of growing awareness and fear concerning the disappearance of humankind's common heritage, selection and assigning value, universalism versus nationalism and particularism, and the dominant Western eye. The statement also reveals how ICOMOS, right from the beginning, considered that outstanding universal value could be discovered beyond the established canons of architecture, if only the methods of modern science were employed.

UNESCO's mission is to save heritage of worldwide interest for posterity. The international protection in the framework of the World Heritage List is founded on the concept of outstanding universal value. States are urged to crystallize this value when presenting nomination dossiers, which are tested and weighed by ICOMOS and IUCN in the (e)valuations that these organizations compile. At the end of these actions the outstanding universal value is confirmed by the World Heritage Committee through its official decision-making process. The impartiality and the scientific and technical nature of the process has been frequently highlighted in official affiliations in order to remove any doubts about its legitimacy. There is no question about UNESCO's and ICOMOS' high professional standards, or the function of the "discourse of objectivity within the system";[2] this, however, does not mean that the concept of outstanding universal value should remain outside the sphere of critical scrutiny. The relevant question is not so much whether right or wrong choices have been made during the nomination and inscription process, but rather what the nature of the process itself is: what vision of World Heritage, and of a World Heritage city in particular, has been constituted and transmitted. Key to this line of inquiry is an understanding of heritage as a social construction and a discursive practice.

Pre-structured objectivity discourse

The period investigated in this research, which covers years from the 1970s until 2011, can without a doubt be labeled a transformation period within the entire heritage sector: both the growth of that sector and its conceptual evolvement are clearly visible. The World Heritage practice in the beginning of the twenty-first century is also very different from the one which was operative in the late 1970s. Since the early 1990s a forceful rhetoric of paradigmatic change has represented a distinctive feature of the official World Heritage discourses. Indeed, many new conceptual openings have been made since that date, for example, those related to the cultural landscapes approach, the definition of authenticity or that of intangible heritage.

This book, by discussing the construction of outstanding universal value for cities has, nevertheless, presented a picture of pre- and post-mid-1990s change, which is more diverse than the official World Heritage rhetoric suggests. On the one hand, in certain areas, for example in relation to reconstruction, the pre-1990s ICOMOS evaluation texts represented ambivalent discursive constructions that already hosted more flexible interpretations. On the other hand, as there have been many reforms since the early/mid-1990s, there have also been many continuities and stabilities. In other words, while there have been attempts to integrate new concepts and to promote pluralization of value as part of the general discourses on World Heritage, many of these attempts have remained largely uninstrumentalized in the actual valuation practice. It has proven difficult for ICOMOS to go beyond the dominant discourse. Therefore, it seems warranted to ask whether some of UNESCO's and ICOMOS' agendas of heritage pluralism have actually been "about *retaining* control in the face of political hostility," that is, not wanting to give up any authority outside the official realm of heritage.[3] While the reform of World Heritage practice over the years is not in question here,[4] it should nevertheless be pointed out how the reform discourse also serves the objective of bringing purpose and legitimacy to support the institution's agenda.

As pointed out several times in this book, a uniform language has characterized the representation and valorization of cities in ICOMOS evaluations. Even though each city is investigated in line with its own past, the developmental trajectories of different cities are presented in a very formal way, making the cities, which are the focus of the description, appear much alike. This is paradoxical in the sense that cities often get engaged in the World Heritage nomination process with quite a different hope in mind – that distinction vis-à-vis other cities would follow the inscription. The uniform language also works to conceal the contested and negotiated nature of heritage.[5]

The ICOMOS evaluation documents' language is influenced by the standardized process of nomination, evaluation and designation. When aiming to create certain global standards, international organizations are inclined to use uniform descriptions to help them categorize a highly diverse group of local places.[6] The evaluation documents obviously need to be structured in a predictable way, as there is a strong impetus for the separate nominations and evaluations to be comparable to each other. Nevertheless, a question remains: how unified and how pre-structured should these descriptions be without obscuring and losing the specific senses of the places? The World Heritage regime has acknowledged the need to involve local communities and to integrate local values and practices into the management of World Heritage sites. But, at the same time, the understanding of outstanding universal value as something separate, even opposite to the values attached to a place by local groups, is ever prevalent in the discourse. This is in line with the argument presented by Waterton, Smith and Campbell concerning the absence of "the idea that conservation values of experts might be just another set of cultural values" in the discursive construction of official heritage texts.[7] While there has been a growing acknowledgement within the World Heritage framework of the position that heritage values are a social construct of time and place, ICOMOS has often treated outstanding universal value as an intrinsic expert value, not least by referring to the places it evaluates as "properties." This perception has been further supported in the ICOMOS evaluation documents by the anonymity of the evaluators, minimal dialogicality and legitimizing techniques that work to validate expert authority. All give the impression of an inevitable, apolitical and technical process devoid of human involvement.

The recent undertaking to publicize the names of the yearly World Heritage Panel members[8] admittedly parts this curtain of apolitical technicality. The growing transparency in the processes of ICOMOS evaluation is not the only issue indicating that the organization has come to a certain kind of crossroads with regard to its involvement in the implementation of the World Heritage Convention. The 2011 World Heritage Committee meeting was significant in that it called the objectivity of ICOMOS into question, as the Committee overruled some recommendations made by ICOMOS concerning the deferral or referral back of certain non-European nominations, such as that of Historic Bridgetown and its Garrison (Barbados). On the other hand, it ought to be remembered that even though relatively rare, this was not the first time that the Committee had overruled ICOMOS' recommendations. For instance, the inclusion of Lübeck in the World Heritage list in 1987 took place despite ICOMOS' negative position. What is new in the current situation is that the debate now concerns European versus non-European heritage values, that is, what kind of emphasis should be placed on material authenticity versus intangible values. Even though the main difference of opinion is perhaps between ICOMOS and the World Heritage Committee, in its own evaluation texts

ICOMOS is also constantly balancing between the two points of the value and authenticity matrix.

In its negative statements, ICOMOS has often referred to a lack of material evidence in support of the significant history of the city and the claims for outstanding universal value, an approach more recently combined under the intermingled considerations of authenticity and integrity. There has recently been a significant change in how ICOMOS presents its negative assessments. While it continues to use "great national value" as an antithesis to outstanding universal value, thus supporting the dominant national narratives associated with World Heritage, the vagueness in the definitions of the early years has been replaced by more systematic reflection with respect to the criteria, authenticity, integrity and comparative frameworks. There is no single answer to the question of how the line has been drawn between outstanding universal value and other value. That outstanding universal value of a once-rejected site may be established after a thematic re-formulation or a new comparative study shows that the line has sometimes been rather fine. In terms of which histories and whose have been narrated and how, there has not been a significant difference noticeable as to whether the document in question proposes acceptance or rejection of the nomination. The history discipline's interest in alternative histories and histories from below, as articulated for example in the context of new social history, gender history, urban history, micro history or spatially informed cultural history, has not been significantly reflected in the ICOMOS narrative of World Heritage cities, even though the narrative has been markedly influenced by the views of global and transnational histories.

Undoubtedly, some of the 'blame' for the restricted scope involving the valuation of cities identified in this research can be placed on national governments. Whenever there is a conservative nomination by a state that is based on traditional criteria and descriptions, the ICOMOS evaluation and the statement of outstanding universal value also tend to be more conventional. Many similarities can be found with the value basis articulated in national World Heritage nomination dossiers.[9] The arguments suggesting that one needs to influence the conservative national nomination policies in order to create a more representative World Heritage List are valid ones,[10] but perhaps they let the international stakeholders, the World Heritage Committee and ICOMOS, off the hook too easily. The international stakeholders are also very responsible for the present contents of the World Heritage List, its value basis and the discursive construction of heritage. ICOMOS cannot reformulate the entire focus of a national nomination it recommends to be rejected. In such a case, it can only ask for a re-nomination, which means that the organization cannot disconnect itself from the national narrative altogether. As ICOMOS is nevertheless an independent agent in the World Heritage valuation process and free to decide on its own narrative, "the capacity for innovation" in the system is not the exclusive right of national governments.[11] As I have demonstrated, ICOMOS has re-articulated, at certain points, the interpretations given in the national nomination document. Ever since the early implementation of the World Heritage Convention, the World Heritage Committee has expected the advisory bodies to be strict in their evaluations of the level of outstanding universal value. Being strict could also mean requiring the States Parties to truly reflect on multiple values, various authenticities and diversified pasts in the nomination dossiers and in reference to all cities alike.

ICOMOS evaluations and the "authorized heritage discourse"

The ICOMOS valuation process lies at the heart of the existence, operation, and power of an internationally working authorized heritage discourse (AHD), as defined by Laurajane Smith.[12] The World Heritage Convention was drafted in conformity with this discourse, and has so far largely remained within its parameters. In addition to privileging expert judgement and the idea of intrinsic value, the articulation of outstanding universal value of cities has also favored monumentality, visual aesthetics, stability and material authenticity. The analysis of ICOMOS urban evaluations has made explicit that the dominant conceptions of heritage are slow to adjust, and that change is slowest at the level of heritage practice. The change can also be only partial. In reference to several World Heritage cities the new conceptual openings were reduced to the level of repeating certain claims, such as, for example, that a site's authenticity has been well maintained, or that a particular city reflects a rich intangible heritage, without further elaboration on what was meant in the particular local context. These kinds of statements further demonstrate how challenging it is to elaborate on the less-traditional heritage aspects and the pluralization of value.

Marc Askew criticizes Smith's notion of authorized heritage discourse for being a "monolithic model" that downplays the important change in the system. Moreover, he argues that Smith's "postcolonial myth of European dominance is belied by the dominance of non-Western practices listed as intangible heritage, as well as the transformation of the World Heritage List itself and the influence of non-Westerners in key positions in its bureaucracy."[13] The examples such as the deletion of Dresden Elbe Valley from the World Heritage List in 2009 or the inclusion of Bridgetown and its Garrison on the List in 2011, both cases in which the non-Western members of the World Heritage Committee took a prominent stand, clearly point to the recent highlighted role of non-Westerners in the World Heritage bureaucracy – something which would definitely warrant more research. It is equally obvious, as shown by many authors, that UNESCO structures and the compiling of the World Heritage List, as well as the whole notion of what constitutes cultural heritage within this framework, have undergone a significant change over the years. Smith too acknowledges the mutability of heritage discourses; she, however, reminds us that the AHD is powerful "because it is a form of 'heritage' itself."[14] In this book I have shown that heritage discourses are susceptible to both stability and change, and that they, at any given time, do not have clearly demarcated boundaries.

With regard to the national nomination files, Sophia Labadi raises the question of whether the non-European nomination dossiers "represent an ambivalent space maintaining but also transgressing the dominant European concepts of heritage." Utilizing the concept of "reiterative universalism," she concludes that this is indeed the case, and that "some of the instances of copying should in no way be considered a submission of non-European countries to Europe," but that they represent deliberately chosen strategies of opposition in the system.[15] While the World Heritage Convention is not a neo-colonialist system,[16] relatively little of this transgressing of the dominant European concepts of heritage and conservation can be found in ICOMOS evaluation texts. Furthermore, it seems that States Parties, Western or non-Western, when nominating new sites to the World Heritage List, define outstanding universal value primarily based upon what they think they are expected to write in order to compile successful proposals.

I would also raise a question about the claim that the transformation of the World Heritage List alone confirms the point concerning a significant change.[17] Even though the transformation of the geographical and thematic scope of urban inscriptions on the World Heritage List during the first decade of the twenty-first century is significant, and clearly indicates a transformation with respect to the valuation of World Heritage cities, it is not the only relevant consideration, as can be seen from the ICOMOS evaluation documents. It appears that new kinds of sites can be constructed in old ways.

World Heritage framing local heritage

Because of the long-time international interchange of ideas and best practices in the field of cultural heritage conservation, it is impossible to pinpoint exactly all of the influences that World Heritage–related concepts have locally. International agendas get absorbed into national discourses and practices and, via them, also into their local counterparts. This is how the AHD is supposed to work: it frames heritage in ways that have implications at the lower levels of heritage practice. At the same time, it is equally clear that implementing international conventions at the national and local levels can lead to manifold interpretations, not always in congruence with the original ideas. Thus, AHD should not be seen as monolithic; it is rather the case that the narrative of place develops "from dialogue including powerful and articulate local voices."[18] The case of Old Rauma as a World Heritage site serves as a testimony to this. While World Heritage has had significant impact on how Old Rauma is talked about and managed as heritage, the Old Rauma case also makes obvious the power of a local narrative of a place and heritage. International definitions and concepts do not automatically, and without modification, travel to and become accepted as part of the local definition of a place. In fact, the most notable feature, if we think about the meeting of the international and local discourses on World Heritage within the context of Old Rauma, is the fact that these two barely met. The case of Old Rauma shows that in the localized context, outstanding universal value gets interpreted in a novel way, sometimes disconnected from the official statements issued. ICOMOS' definition of a place – Old Rauma as a conservatory of traditional settlements – did not prove to be influential, mostly because this statement of outstanding universal value was not shared locally. In Rauma, World Heritage and the concept of outstanding universal value were used as rhetorical tools in the construction of heritage narratives of local origin; however, these notions and their use were significantly influenced by the prior narratives and meanings associated with Old Rauma. Old Rauma's image as a 'lively wooden town' was produced decades before the World Heritage status, and reproduced – one way or another – anew on almost every occasion when the area's past and future were (re)considered.

Therefore, in the world of cultural globalization and transnational memory work, and in opposition to certain other examples such as Vienna and Dresden, Old Rauma provides a locally and nationally framed example. From the point of view of local involvement in heritage this seems empowering, although, at the same time, discouraging from the point of view of international 'ownership' and the control of potentially harmful impacts that urban development projects have on World Heritage cities. Not all locals necessarily have a very deep understanding of what the World Heritage system or the outstanding universal value are about, but the example of Old Rauma suggests that they have a firm grip on their heritage. Local people's participation in the management of World Heritage sites is currently something that is strongly encouraged by the World Heritage Committee and ICOMOS. What these organizations need to anticipate is that the local participation does not always have positive outcomes in light of their own objectives.

In Old Rauma, the local community's firm grip on their heritage did not mean the existence of a unanimously shared local narrative. Even in a rather stable World Heritage setting, conflicts over meaning and use can arise. In Old Rauma's case several understandings of outstanding universal value were in effect locally, with visual and social/functional understandings representing the opposite poles. This shows the importance of communicating the statement of outstanding universal value when it has been adopted. For those in Old Rauma approaching the concept of outstanding universal value as a merely visual category, any considerations beyond visual effects on the World Heritage area had virtually nothing to do with maintaining or losing outstanding universal value. Those groups opposing the construction of a shopping center on the Old Rauma buffer zone understood outstanding universal value to be based on social and functional values and continuities. These intangible values proved to be difficult to endorse when they came into conflict with other values and interests. The case of Old Rauma also shows the mutability of the "arena of heritage conflict."[19] For a long time the most visible conflict over the meaning of Old Rauma was between local conservation authorities and the Old Rauma shopkeepers. In the situation arising from the shopping center debate, however, these old opponents joined forces to oppose outside enemies. Even though unsuccessful in their ultimate goal of blocking the building of the shopping center, the shopkeepers were successful in translating the commercial liveliness of Old Rauma into a new language, having it accepted by other interest groups as part of World Heritage.

A future for urban World Heritage

This book has offered a historical analysis focusing on the transnational application so far of the concept of outstanding universal value in the context of cities. However, a research topic such as this, which comes up to the present and touches upon ongoing discussions, seems to also require some commentary concerning future action. With this in mind, I would now like to make three comments concerning the future implementation of the World Heritage Convention. Firstly, it seems highly recommendable to critically reconsider the three categories of 'monuments,' 'groups of buildings' and 'sites,' which are offered as a definition for cultural heritage in the World Heritage Convention. Perhaps a future strategy could be to replace the three categories by introducing a common denominator, such as 'place,' for various types of cultural heritage (as in the Burra Charter, the principles of which, even though "generally accepted at the centre,"[20] remain unspecified in the context of World Heritage). This conceptual change could lead to more freely elaborating on the different aspects of past and present, value and outstanding universal value within the World Heritage framework. Or, if the three categories were not to be eliminated altogether, as this would mean undertaking the burdensome process of redrafting of the World Heritage Convention text, at least there should be a more flexible use of the existing categories. By this I refer primarily to the category of 'sites' and the Operational Guidelines' category of 'cultural landscapes,' both of which could offer a wider potential to facilitate more flexible conceptualizations of urban heritage and its conservation and of heritage values. As argued throughout this book, it seems obvious that the category so far most often applied within the context of cities, the 'groups of buildings,' is clearly outdated in its emphasis on the features related to architecture, homogeneity, unity and stability. The application of the 'groups of buildings' category, especially its embedded notion of townscape harmony, has created a rather monotonous way of thinking about how urban heritage can and should be assessed, and about which values are relevant within its context. The concept of a 'historic urban landscape' has the potential to solve some of these discrepancies – and a renewal of the terminology has indeed been proposed. But since no significant changes to the Operational Guidelines have ensued so far from this conceptual opening, it remains unclear what kind of role that opening will play in the World Heritage valuation process.

My second suggestion involves reviewing the criteria used in the determination of outstanding universal value. Even though concern for the representation of different cultures and geographical areas remains without a doubt key to the future implementation of the Convention, there should also be a growing concern for breaking with the old conventions of representation of World Heritage cities and other sites. The heterogeneity of urban heritage and urban values should be truly acknowledged, also in regard to Western urban heritage. A radical suggestion could even be to remove the predetermined criteria from the evaluation process altogether, since they are inclined to produce similar descriptions. It is interesting to note that the actual utilization of the criteria in the context of urban heritage has recently been developing towards a more uniform use: since 2002 criterion i has been used only once in reference to cities, and criteria v and vi only occasionally. Furthermore, in its recent decisions ICOMOS has indicated that the application of criterion ii should seldom occur.[21] In other words, most future urban nominations should fall under criteria iii and iv. Another visible trend in the recent ICOMOS evaluations has been their increasing concentration on comparative assessments – something which indicates that the criteria have a declining importance. In any case, the criteria would benefit from renewal towards a practice which would allow them to name more explicitly the different heritage values. This explicit naming of values in the criteria could encourage states to adopt a similar kind of practice in their nomination dossiers, and could contribute to considering the different values more equal in the valuation and management of World Heritage cities. Up until now, and despite the seeming recent multivalence, values other than the historical and the aesthetic have been given a secondary status. The argument that there is a depoliticized "international vocabulary of cultural and artistic essences" embedded in the World Heritage criteria still seems valid.[22]

The question of development within the conserved city has been the key concern for an integrated conservation approach since the 1970s. Within the World Heritage framework the question concerning managing change in historic urban environments became intensified at the turn of the new millennium as part of the process leading up to the adoption of the Vienna Memorandum in 2005, and, in 2011, the UNESCO Recommendation on the Historic Urban Landscape. Despite these discussions, clearly marking new ways in how UNESCO and ICOMOS approach historic cities, they have largely been excluded from the parameters of World Heritage valuation. The ICOMOS evaluations have, up until now, relied on the definition of material permanency of the heritage 'resource' as it has developed so far, and conversely paid less attention to any such (positive) processes of change that might be characteristic of the city. The third point for future reflection that I wish to raise is that perhaps they should.

The ICOMOS valuation documents leave the history of the twentieth century almost systematically with very little mention. The counterargument here of course could be that these newer elements rarely represent outstanding universal value. Similarly, it might be objected that change and development are brought into focus in the context of World Heritage cities in the first place. By bringing these aspects into focus, is there a danger of harming conservation's agenda? In my view the risk is worth taking. By stating this I do not wish to argue that 'anything goes' in World Heritage cities. Nor do I wish to argue that change in these places should somehow be an end in itself. Rather, I have argued in this book in favor of the importance of stating urban continuity as a process – in all its diversity – and in favor of an even broader inclusion of social values and 'living' urban heritage in the statements of outstanding universal value of cities. In an increasingly urbanizing world the question of urban heritage seems as important as ever.

Notes

1 UNESCO, Informal Consultation of Intergovernmental and non-Governmental Organizations on the Implementation of the Convention Concerning the Protection of the World Cultural and Natural Heritage. Morges, 19–20 May 1976. CC-76/WS/25. Annex III, Proposals made by the International Council on Monuments and Sites, p. 2.

UNESCO, World Heritage Centre, WH Committee & Bureau, Working Documents from 1976 to 1987 (CD in author's possession).

2 Long and Labadi, "Introduction," 8–9.

3 Gibson and Pendlebury, "Introduction: Valuing Historic Environments," 10. Original emphasis.

4 Cameron and Rössler, *Many Voices*.

5 Rico, "Negative heritage," 349.

6 See also Green, *Managing Laponia*, 85.

7 Waterton, Smith and Campbell, "The Utility of Discourse Analysis," 349.

8 See, for example, ICOMOS News: ICOMOS World Heritage Panel 2014, accessed February 2, 2016, www.icomos.org/images/DOCUMENTS/World_Heritage/ICOMOS_World_Heritage_Panel_2014.pdf.

9 Labadi, "Representations of the Nation"; Labadi, *UNESCO*.

10 Cleere, "The Uneasy Bedfellows"; Labadi, "A Review"; Labadi, "Representations of the Nation."

11 Cf. Askew, "The Magic List," 32.

12 Smith, *Uses of Heritage*.

13 Askew, "The Magic List," 32.

14 Smith, *Uses of Heritage*, 299.

15 Labadi, *UNESCO*, 21, 152.

16 Ibid., 21.

17 Askew, "The Magic List," 32.

18 Gibson and Pendlebury, "Introduction: Valuing Historic Environments," 10; John Pendlebury, Tim Townshend and Rose Gilroy, "Social Housing as Heritage: The Case of Byker, Newcastle upon Tyne," in *Valuing Historic Environments*, ed. Lisanne Gibson and John Pendlebury (Surrey: Ashgate, 2009), 199.

19 Smith, *Uses of Heritage*, 187.

20 Logan, "Globalizing Heritage," 56.

21 See, for example, ICOMOS evaluation for the nomination of the World Heritage property, "La Chaux-de-Fonds/Le Locle," Switzerland, No 1302, 10 March 2009.

22 Hevia, "World Heritage, National Culture," 224. See also Askew, "The Magic List," 34.

Appendices

Appendix 1: World Heritage cities (with their official full titles) by year, state and UNESCO Region

Year	World Heritage city	State	UNESCO Region
1978	City of **Quito**	Ecuador	Lat. Am./Carib.
	Historic Centre of **Cracow**	Poland	Eur./N-A
1979	Ancient City of **Damascus**	Syrian Arab Republic	Arab states
	Antigua Guatemala	Guatemala	Lat. Am./Carib.
	Bryggen	Norway	Eur./N-A
	Historical Complex of **Split** with the Palace of Diocletian	Croatia	Eur./N-A
	Historic **Cairo**	Egypt	Arab states
	Medina of **Tunis**	Tunisia	Arab states
	Natural and Culturo-Historical Region of **Kotor**	Montenegro	Eur./N-A
	Old City of **Dubrovnik**	Croatia	Eur./N-A
1980	City of **Valletta**	Malta	Eur./N-A
	Historic Centre of **Rome**, the Properties of the Holy See in that City Enjoying Extraterritorial Rights and San Paolo Fuori le Mura	Italy	Eur./N-A
	Historic Centre of **Warsaw**	Poland	Eur./N-A
	Historic Town of **Ouro Preto**	Brazil	Lat. Am./Carib.

(continued)

Year	World Heritage city	State	UNESCO Region
	Natural and Cultural Heritage of the **Ohrid** Region	The former Yugoslav Republic of Macedonia	Eur./N-A
	Røros Mining Town and the Circumference	Norway	Eur./N-A
1981	Medina of **Fez**	Morocco	Arab states
	Old City of **Jerusalem** and its Walls	Jerusalem (nom. by Jordan)	Eur./N-A
1982	Historic Centre of **Florence**	Italy	Eur./N-A
	Historic Centre of the Town of **Olinda**	Brazil	Lat. Am./Carib.
	M'Zab Valley	Algeria	Arab states
	Old **Havana** and its Fortification System	Cuba	Lat. Am./Carib.
	Old Walled City of **Shibam**	Yemen	Arab states
1983	Ancient City of **Nessebar**	Bulgaria	Eur./N-A
	Central Zone of the Town of **Angra do Heroismo** in the Azores	Portugal	Eur./N-A
	City of **Cuzco**	Peru	Lat. Am./Carib.
	Old City of **Berne**	Switzerland	Eur./N-A
1984	Port, Fortresses and Group of Monuments, **Cartagena**	Colombia	Lat. Am./Carib.
1985	Historic Areas of **Istanbul**	Turkey	Eur./N-A
	Historic Centre of **Salvador de Bahia**	Brazil	Lat. Am./Carib.
	Historic District of Old **Quebec**	Canada	Eur./N-A
	Medina of **Marrakesh**	Morocco	Arab states
	Old Town of **Ávila** with its Extra-Muros Churches	Spain	Eur./N-A
	Old Town of **Segovia** and its Aqueduct	Spain	Eur./N-A
	Santiago de Compostela (Old Town)	Spain	Eur./N-A

Year	World Heritage city	State	UNESCO Region
1986	Ancient City of **Aleppo**	Syrian Arab Republic	Arab states
	Historic Centre of **Évora**	Portucal	Eur./N-A
	Historic City of **Toledo**	Spain	Eur./N-A
	Old City of **Sana'a**	Yemen	Arab states
	Old Town of **Cáceres**	Spain	Eur./N-A
	Old Town of **Ghadamès**	Libyan Arab Jamahiriya	Arab states
1987	**Brasilia**	Brazil	Lat. Am./Carib.
	Budapest, including the Banks of the Danube, the Buda Castle Quarter and Andrássy Avenue	Hungary	Eur./N-A
	City of **Bath**	UK	Eur./N-A
	City of **Potosi**	Bolivia	Lat. Am./Carib.
	Hanseatic City of **Lübeck**	Germany	Eur./N-A
	Historic Centre of **Mexico City** and Xochimilco	Mexico	Lat. Am./Carib.
	Historic Centre of **Oaxaca** and Archaeological Site of Monte Albán	Mexico	Lat. Am./Carib.
	Historic Centre of **Puebla**	Mexico	Lat. Am./Carib.
	Venice and its Lagoon	Italy	Eur./N-A
1988	Historic Town of **Guanajuato** and Adjacent Mines	Mexico	Lat. Am./Carib.
	Kairouan	Tunisia	Arab states
	Medieval City of **Rhodes**	Greece	Eur./N-A
	Medina of **Sousse**	Tunisia	Arab states
	Old City of **Salamanca**	Spain	Eur./N-A
	Old Town of **Galle** and its Fortifications	Sri Lanka	Asia/Pacific
	Old Town of **Djenné**	Mali	Africa
	Strasbourg – Grande île	France	Eur./N-A

(continued)

Appendix 1 (*continued*)

Year	World Heritage city	State	UNESCO Region
	Timbuktu	Mali	Africa
	Trinidad and the Valley de los Ingenios	Cuba	Lat. Am./Carib.
1990	Colonial City of **Santo Domingo**	Dominican Republic	Lat. Am./Carib.
	Historic Centre of **Saint Petersburg** and Related Groups of Monuments	Russian Federation	Eur./N-A
	Historic Centre of **San Gimignano**	Italy	Eur./N-A
	Historic Centre of **Lima**	Peru	Lat. Am./Carib.
1991	Historic Centre of **Morelia**	Mexico	Lat. Am./Carib.
	Historic City of **Sucre**	Bolivia	Lat. Am./Carib.
	Island of **Mozambique**	Mozambique	Africa
	Old **Rauma**	Finland	Eur./N-A
	Paris, Banks of the Seine	France	Eur./N-A
1992	Historic Centre of **Český Krumlov**	Czech Republic	Eur./N-A
	Historic Centre of **Prague**	Czech Republic	Eur./N-A
	Historic Centre of **Telč**	Czech Republic	Eur./N-A
	Kasbah of **Algiers**	Algeria	Arab states
	Mines of Rammelsberg, Historic Town of **Goslar** and Upper Harz Water Management System	Germany	Eur./N-A
	Old City of **Zamość**	Poland	Eur./N-A
1993	**Coro** and its Port	Venezuela	Lat. Am./Carib.
	Historic Centre of **Bukhara**	Uzbekistan	Asia/Pacific
	Historic Centre of **Zacatecas**	Mexico	Lat. Am./Carib.
	Historic Town of **Banská Štiavnica** and the Technical Monuments in its Vicinity	Slovakia	Eur./N-A
	Historic Town of **Zabid**	Yemen	Arab states
	The Sassi and the park of the Rupestrian Churches of **Matera**	Italy	Eur./N-A

Year	World Heritage city	State	UNESCO Region
	Town of **Bamberg**	Germany	Eur./N-A
1994	Alhambra, Generalife and Albayzin, **Granada**	Spain	Eur./N-A
	City of **Luxembourg**: its Old Quarters and Fortifications	Luxembourg	Eur./N-A
	City of **Safranbolu**	Turkey	Eur./N-A
	City of **Vicenza** and the Palladian Villas of the Veneto	Italy	Eur./N-A
	Collegiate Church, Castle, and Old Town of **Quedlinburg**	Germany	Eur./N-A
	Historic Centre of **Cordoba**	Spain	Eur./N-A
	Vilnius Historic Centre	Lithuania	Eur./N-A
1995	Cultural Landscape of **Sintra**	Portugal	Eur./N-A
	Ferrara, City of the Renaissance and its Po Delta	Italy	Eur./N-A
	Hanseatic Town of **Visby**	Sweden	Eur./N-A
	Historic Centre of **Naples**	Italy	Eur./N-A
	Historic Centre of **Santa Cruz de Mompox**	Colombia	Lat. Am./Carib.
	Historic Centre of **Siena**	Italy	Eur./N-A
	Historic Quarter of the City of **Colonia del Sacramento**	Uruguay	Lat. Am./Carib.
	Kutná Hora: Historical Town Centre with the Church of St Barbara and the Cathedral of Our Lady at Sedlec	Czech Republic	Eur./N-A
	Old and New Towns of **Edinburgh**	UK	Eur./N-A
	Old Town **Lunenburg**	Canada	Eur./N-A
	Town of **Luang Prabang**	Lao People's Democratic Republic	Asia/Pacific
1996	Ancient Ksour of **Ouadane, Chinguetti, Tichitt** and **Oualata**	Mauritania	Arab states
	Historic Centre of the City of **Pienza**	Italy	Eur./N-A

(continued)

Appendix 1 (*continued*)

Year	World Heritage city	State	UNESCO Region
	Historic Centre of the City of **Salzburg**	Austria	Eur./N-A
	Historic Centre of **Oporto**, Luiz I Bridge and Monastery of Serra do Pilar	Portugal	Eur./N-A
	Historic City of **Meknes**	Morocco	Arab states
	Historic Monuments Zone of **Querétaro**	Mexico	Lat. Am./Carib.
	Historic Walled Town of **Cuenca**	Spain	Eur./N-A
1997	Ancient City of **Ping Yao**	China	Asia/Pacific
	Archaeological Site of Panamá Viejo and Historic District of **Panamá**	Panama	Lat. Am./Carib.
	Hallstatt-Dachstein Salzkammergut Cultural Landscape	Austria	Eur./N-A
	Historic Area of **Willemstad**, Inner City and Harbour, Netherlands Antilles	Netherlands	Eur./N-A
	Historic Centre (Old Town) of **Tallinn**	Estonia	Eur./N-A
	Historic Centre of **Riga**	Latvia	Eur./N-A
	Historic Centre of **São Luis**	Brazil	Lat. Am./Carib.
	Historic City of **Trogir**	Croatia	Eur./N-A
	Historic Fortified City of **Carcassonne**	France	Eur./N-A
	Maritime **Greenwich**	UK	Eur./N-A
	Medieval Town of **Torun**	Poland	Eur./N-A
	Medina of **Tétouan** (formerly known as Titawin)	Morocco	Arab states
	Old Town of **Lijiang**	China	Asia/Pacific
	Portovenere, Cinque Terre, and the Islands (Palmaria, Tino and Tinetto)	Italy	Eur./N-A
1998	Historic Centre of **Urbino**	Italy	Eur./N-A
	Historic Monuments Zone of **Tlacotalpan**	Mexico	Lat. Am./Carib.
	Historic Site of **Lyons**	France	Eur./N-A
	L'viv – the Ensemble of the Historic Centre	Ukraine	Eur./N-A

Year	World Heritage city	State	UNESCO Region
	Naval Port of **Karlskrona**	Sweden	Eur./N-A
	University and Historic Precinct of **Alcalá de Henares**	Spain	Eur./N-A
1999	City of **Graz** – Historic Centre and Schloss Eggenberg	Austria	Eur./N-A
	Historic Centre (Chorá) with the Monastery of Saint John "the Theologian" and the Cave of the Apocalypse on **the Island of Pátmos**	Greece	Eur./N-A
	Historic Centre of **Santa Ana de los Ríos de Cuenca**	Ecuador	Lat. Am./Carib.
	Historic Centre of **Sighişoara**	Romania	Eur./N-A
	Historic Centre of the Town of **Diamantina**	Brazil	Lat. Am./Carib.
	Historic Fortified Town of **Campeche**	Mexico	Lat. Am./Carib.
	Historic Town of **Vigan**	Philippines	Asia/Pacific
	Hoi An Ancient Town	Viet Nam	Asia/Pacific
	Ibiza, biodiversity and culture	Spain	Eur./N-A
	San Cristóbal de La Laguna	Spain	Eur./N-A
2000	**Assisi**, the Basilica of San Francesco and Other Franciscan Sites	Italy	Eur./N-A
	Bardejov Town Conservation Reserve	Slovakia	Eur./N-A
	City of **Verona**	Italy	Eur./N-A
	Historic Centre of **Brugge**	Belgium	Eur./N-A
	Historic Centre of **Shakhrisyabz**	Uzbekistan	Asia/Pacific
	Historical Centre of the City of **Arequipa**	Peru	Lat. Am./Carib.
	Historic Town of **St George** and Related Fortifications, Bermuda	UK (Bermuda)	Eur./N-A
	Island of **Saint-Louis**	Senegal	Africa
	Stone Town of **Zanzibar**	United Republic of Tanzania	Africa
	Walled City of **Baku** with the Shirvanshah's Palace and Maiden Tower	Azerbaijan	Asia/Pacific

(*continued*)

Appendix 1 (*continued*)

Year	World Heritage city	State	UNESCO Region
2001	Historic Centre of **Guimarães**	Portugal	Eur./N-A
	Historic Centre of the Town of **Goiás**	Brazil	Lat. Am./Carib.
	Historic Centre of **Vienna**	Austria	Eur./N-A
	Lamu Old Town	Kenya	Africa
	Medina of **Essaouira** (formerly Mogador)	Morocco	Arab states
	Old City of **Acre**	Israel	Eur./N-A
	Provins, Town of Medieval Fairs	France	Eur./N-A
	Samarkand - Crossroads of Culture	Uzbekistan	Asia/Pacific
2002	Historic Centres of **Stralsund** and **Wismar**	Germany	Eur./N-A
	Historic Inner City of **Paramaribo**	Suriname	Lat. Am./Carib.
	Late Baroque Towns of the Val di Noto (South-Eastern Sicily)	Italy	Eur./N-A
2003	Citadel, Ancient City and Fortress Buildings of **Derbent**	Russian Federation	Eur./N-A
	Historic Quarter of the Seaport City of **Valparaíso**	Chile	Lat. Am./Carib.
	Jewish Quarter and St Procopius' Basilica in **Třebíč**	Czech Republic	Eur./N-A
	Renaissance Monumental Ensembles of **Úbeda** and **Baeza**	Spain	Eur./N-A
	The White City of **Tel-Aviv** - the Modern Movement	Israel	Eur./N-A
2004	**Dresden** Elbe Valley	Germany	Eur./N-A
	Liverpool – Maritime Mercantile City	UK	Eur./N-A
	Portuguese City of **Mazagan (El Jadida)**	Morocco	Arab states
2005	Historic Centre of **Macao**	China	Asia/Pacific
	Historical Centre of the City of **Yaroslavl**	Russian Federation	Eur./N-A
	Le Havre, the City Rebuilt by Auguste Perret	France	Eur./N-A

Year	World Heritage city	State	UNESCO Region
	Historic Centres of **Berat** and **Gjirokastra**, formerly Museum-City of Gjirokastra	Albania	Eur./N-A
	Old Bridge Area of the Old City of **Mostar**	Bosnia and Herzegovina	Eur./N-A
	Urban Historic Centre of **Cienfuegos**	Cuba	Lat. Am./Carib.
2006	**Harar Jugol**, the Fortified Historic Town	Ethiopia	Africa
	Old Town of **Regensburg** with Stadtamhof	Germany	Eur./N-A
2007	**Bordeaux**, Port of the Moon	France	Eur./N-A
	Old Town of **Corfu**	Greece	Eur./N-A
2008	Historic Centre of **Camagüey**	Cuba	Eur./N-A
	Berlin Modernism Housing Estates	Germany	Eur./N-A
	Mantua and **Sabbioneta**	Italy	Eur./N-A
	Melaka and **George Town**, Historic Cities of the Straits of Malacca	Malaysia	Asia/Pacific
	Protective town of **San Miguel** and the Sanctuary of Jesús Nazareno de Atotonilco	Mexico	Lat. Am./Carib.
	San Marino Historic Centre and Mount Titano	San Marino	Eur./N-A
2009	**Cidade Velha**, Historic Centre of Ribeira Grande	Cape Verde	Africa
	La Chaux-de-Fonds/Le Locle, Watchmaking Town Planning	Switzerland	Eur./N-A
2010	Episcopal City of **Albi**	France	Eur./N-A
	Seventeenth-Century Canal Ring Area of **Amsterdam** inside the Singelgracht	The Netherlands	Eur./N-A
2011	**Bridgetown** and its Garrison	Barbados	Lat. Am./Carib.

Appendix 2: Rejected urban World Heritage nominations

Urban area subject to nomination	Nominating State Party	Date of rejection (date of subsequent inscription)
Old Stone Town of Zanzibar	Tanzania	1982 (2000)
Ancient City of Plovdiv	Bulgaria	1983 (–)
Hanseatic City of Lübeck	Germany	1983 (1987)
Tripoli	Lebanon	1984 (–)
Bascarsija, the Historic Centre of the City of Sarajevo	Yugoslavia	1986 (–)
Sur al-Luwatiya, the Historic Centre of Matrah	Oman	1987 (–)
Dalt Vila (Ibiza)	Spain	1987 (incl. in 1999 as a cultural landscape "Ibiza, biodiversity and culture")
Port Royal	Jamaica	1988 (–)
Historic Centre of Popayán	Colombia	1989 (–)
Gerona	Spain	1989 (–)
Úbeda and Baeza	Spain	1989 (2003)
Town of Taal	Philippines	1989 (–)
Town of Vigan	Philippines	1989 (1999)
Historic Centre Manila Intramuros	Philippines	1989 (–)
Dresden (Baroque ensemble)	GDR / Germany	1990 (incl. in 2004 as a cultural landscape "Dresden Elbe Valley"; deleted in 2009)
Historic Centre of Riga	USSR / Latvia	1991 (1997)
Cidade Velha	Cape Verde	1992 (2009)
Savannah City Plan	USA	1995 (–)
Medina of Essaouira	Morocco	1996
The Medieval Town of Provins	France	1998 (2001)
Cultural Stratification in the Historic Centre of the City of Pecs	Hungary	1998 (incl. in 2000 as monumental complex "Early Christian Necropolis of Pécs (Sopianae)")

Urban area subject to nomination	Nominating State Party	Date of rejection (date of subsequent inscription)
Gdańsk: The Main Town, the Motlava Side Channel, and the Vistula Mouth Fortress	Poland	1998 (–)
Sarajevo – Unique symbol of universal multiculture and continual open city	Bosnia and Herzegovina	1999 (–)
City of La Plata, Foundational Urban Plan	Argentina	2000 (–)
Valparaíso	Chile	2000 (2003)
Historic Centre of Santa Fe de Bogotá	Colombia	2000 (–)
Old Town of Corfu	Greece	2000 (2007)
Historic Centre of Santarém	Portugal	2000 (–)
The Renaissance Monumental Ensembles of Úbeda and Baeza	Spain	2000 (2003)
Historic Centre of Trujillo	Peru	2002 (–)
Historic City of Mardin	Turkey	2003 (–)
Historic Centre of Innsbruck with Schloss Ambras and Nordkette-Karwendel Alpine Park	Austria	2005 (–)
Trakai Historical National Park	Lithuania	2005 (–)
Ancient City of Plovdiv	Bulgaria	2006 (–)
Foundational City Area of La Plata	Argentina	2007 (–)
Gdańsk – The Site of Memory and Freedom	Poland	2007 (–)
Sibiu, the Historic Centre	Romania	2007 (–)
Cultural Landscape of Buenos Aires	Argentina	2008 (–)
Cultural Property of the Historic Town of Jaice	Bosnia and Herzegovina	2009 (–)
Historical City of Jeddah	Saudi Arabia	2011 (–)
Old City and Ramparts of Alanya	Turkey	2011 (–)

Appendix 3: The use of cultural heritage criteria in reference to cities

World Heritage city	(i)	(ii)	(iii)	(iv)	(v)	(vi)
City of Quito		X		X		
Historic Centre of Cracow				X		
Ancient City of Damascus	X	X	X	X		X
Antigua Guatemala		X	X	X		
Bryggen			X			
Historical Complex of Split with the Palace of Diocletian		X	X	X		
Historic Cairo	X				X	X
Medina of Tunis		X	X		X	
Natural and Culturo-Historical Region of Kotor	X	X	X	X		
Old City of Dubrovnik	X		X	X		
City of Valletta	X					X
Historic Centre of Rome, the Properties of the Holy See in that City Enjoying Extraterritorial Rights and San Paolo Fuori le Mura	X	X	X	X		X
Historic Centre of Warsaw		X				X
Historic Town of Ouro Preto	X		X			
Natural and Cultural Heritage of the Ohrid Region	X		X	X		
Røros Mining Town and the Circumference			X	X	X	
Medina of Fez		X			X	
Old City of Jerusalem and its Walls		X	X			X
Historic Centre of Florence	X	X	X	X		X
Historic Centre of the Town of Olinda	X			X		
M'Zab Valley		X	X		X	
Old Havana and its Fortification System			X	X		
Old Walled City of Shibam			X	X	X	

World Heritage city	(i)	(ii)	(iii)	(iv)	(v)	(vi)
Ancient City of Nessebar			X	X		
Central Zone of the Town of Angra do Heroismo in the Azores				X		X
City of Cuzco			X	X		
Old City of Berne			X			
Port, Fortresses and Group of Monuments, Cartagena				X		X
Historic Areas of Istanbul	X	X	X	X		
Historic Centre of Salvador de Bahia				X		X
Historic District of Old Quebec				X		X
Medina of Marrakesh	X	X		X	X	
Old Town of Ávila with its Extra-Muros Churches			X	X		
Old Town of Segovia and its Aqueduct	X		X	X		
Santiago de Compostela (Old Town)	X	X				X
Ancient City of Aleppo			X	X		
Historic Centre of Évora		X		X		
Historic City of Toledo	X	X	X	X		
Old City of Sana'a				X	X	X
Old Town of Cáceres			X	X		
Old Town of Ghadamès					X	
Brasilia	X			X		
Budapest, including the Banks of the Danube, the Buda Castle Quarter and Andrássy Avenue		X		X		
City of Bath	X	X		X		
City of Potosi		X		X		X
Hanseatic City of Lübeck				X		
Historic Centre of Mexico City and Xochimilco		X	X	X	X	
Historic Centre of Oaxaca and Archaeological Site of Monte Albán	X	X	X	X		

(continued)

Appendix 3 (*continued*)

World Heritage city	(i)	(ii)	(iii)	(iv)	(v)	(vi)
Historic Centre of Puebla		X		X		
Venice and its Lagoon	X	X	X	X	X	X
Historic Town of Guanajuato and Adjacent Mines	X	X		X		X
Kairouan	X	X	X		X	X
Medieval City of Rhodes		X		X	X	
Medina of Sousse			X	X	X	
Old City of Salamanca	X	X		X		
Old Town of Galle and its Fortifications				X		
Old Town of Djenné			X	X		
Strasbourg – Grande île	X	X		X		
Timbuktu		X		X	X	
Trinidad and the Valley de los Ingenios				X	X	
Colonial City of Santo Domingo		X		X		X
Historic Centre of Saint Petersburg and Related Groups of Monuments	X	X		X		X
Historic Centre of San Gimignano	X		X	X		
Historic Centre of Lima				X		
Historic Centre of Morelia		X		X		X
Historic City of Sucre				X		
Island of Mozambique				X		X
Old Rauma				X	X	
Paris, Banks of the Seine	X	X		X		
Historic Centre of Český Krumlov				X		
Historic Centre of Prague		X		X		X
Historic Centre of Telč	X			X		
Kasbah of Algiers		X			X	

World Heritage city	(i)	(ii)	(iii)	(iv)	(v)	(vi)
Mines of Rammelsberg, Historic Town of Goslar and Upper Harz Water Management System	X	X	X	X		
Old City of Zamość				X		
Coro and its Port				X	X	
Historic Centre of Bukhara		X		X		X
Historic Centre of Zacatecas		X		X		
Historic Town of Banská Štiavnica and the Technical Monuments in its Vicinity				X	X	
Historic Town of Zabid		X		X		X
The Sassi and the park of the Rupestrian Churches of Matera			X	X	X	
Town of Bamberg		X		X		
Alhambra, Generalife and Albayzin, Granada	X	X		X		
City of Luxembourg: its Old Quarters and Fortifications				X		
City of Safranbolu		X		X	X	
City of Vicenza and the Palladian Villas of the Veneto	X	X				
Collegiate Church, Castle, and Old Town of Quedlinburg				X		
Historic Centre of Cordoba	X	X	X	X		
Vilnius Historic Centre		X		X		
Cultural Landscape of Sintra		X		X	X	
Ferrara, City of the Renaissance and its Po Delta		X	X	X	X	X
Hanseatic Town of Visby				X	X	
Historic Centre of Naples		X		X		
Historic Centre of Santa Cruz de Mompox				X	X	
Historic Centre of Siena	X	X		X		
Historic Quarter of the City of Colonia del Sacramento				X		

(*continued*)

Appendix 3 (*continued*)

World Heritage city	(i)	(ii)	(iii)	(iv)	(v)	(vi)
Kutná Hora: Historical Town Centre with the Church of St Barbara and the Cathedral of Our Lady at Sedlec		X		X		
Old and New Towns of Edinburgh		X		X		
Old Town Lunenburg				X	X	
Town of Luang Prabang		X		X	X	
Ancient Ksour of Ouadane, Chinguetti, Tichitt and Oualata			X	X	X	
Historic Centre of the City of Pienza	X	X		X		
Historic Centre of the City of Salzburg		X		X		X
Historic Centre of Oporto, Luiz I Bridge and Monastery of Serra do Pilar				X		
Historic City of Meknes				X		
Historic Monuments Zone of Querétaro		X		X		
Historic Walled Town of Cuenca		X			X	
Ancient City of Ping Yao		X	X	X		
Archaeological Site of Panamá Viejo and Historic District of Panamá			X	X	X	
Hallstatt-Dachstein Salzkammergut Cultural Landscape			X	X		
Historic Area of Willemstad, Inner City and Harbour, Netherlands Antilles		X		X	X	
Historic Centre (Old Town) of Tallinn		X		X		
Historic Centre of Riga	X	X				
Historic Centre of São Luis			X	X	X	
Historic City of Trogir		X		X		
Historic Fortified City of Carcassonne		X		X		
Maritime Greenwich	X	X		X		X
Medieval Town of Torun		X		X		
Medina of Tétouan (formerly known as Titawin)		X		X	X	

World Heritage city	(i)	(ii)	(iii)	(iv)	(v)	(vi)
Old Town of Lijiang		X		X	X	
Portovenere, Cinque Terre, and the Islands (Palmaria, Tino and Tinetto)	X			X	X	
Historic Centre of Urbino		X		X		
Historic Monuments Zone of Tlacotalpan		X		X		
Historic Site of Lyons		X		X		
L'viv – the Ensemble of the Historic Centre		X			X	
Naval Port of Karlskrona		X		X		
University and Historic Precinct of Alcalá de Henares		X		X		X
City of Graz – Historic Centre and Schloss Eggenberg		X		X		
Historic Centre (Chorá) with the Monastery of Saint John "the Theologian" and the Cave of the Apocalypse on the Island of Pátmos			X	X		X
Historic Centre of Santa Ana de los Ríos de Cuenca		X		X	X	
Historic Centre of Sighişoara			X		X	
Historic Centre of the Town of Diamantina		X		X		
Historic Fortified Town of Campeche		X		X		
Historic Town of Vigan		X		X		
Hoi An Ancient Town		X			X	
Ibiza, biodiversity and culture			X	X	X	
San Cristóbal de La Laguna		X		X		
Assisi, the Basilica of San Francesco and Other Franciscan Sites	X	X	X	X		X
Bardejov Town Conservation Reserve			X	X		
City of Verona		X		X		
Historic Centre of Brugge		X		X		X
Historic Centre of Shakhrisyabz			X	X		
Historical Centre of the City of Arequipa	X			X		

(continued)

Appendix 3 (*continued*)

World Heritage city	(i)	(ii)	(iii)	(iv)	(v)	(vi)
Historic Town of St George and Related Fortifications, Bermuda	X			X		
Island of Saint-Louis		X		X		
Stone Town of Zanzibar		X	X			X
Walled City of Baku with the Shirvanshah's Palace and Maiden Tower				X		
Historic Centre of Guimarães		X	X	X		
Historic Centre of the Town of Goiás		X		X		
Historic Centre of Vienna		X		X		X
Lamu Old Town		X		X		X
Medina of Essaouira (formerly Mogador)		X		X		
Old City of Acre		X	X		X	
Provins, Town of Medieval Fairs		X		X		
Samarkand – Crossroads of Culture	X	X		X		
Historic Centres of Stralsund and Wismar		X		X		
Historic Inner City of Paramaribo		X		X		
Late Baroque Towns of the Val di Noto (South-Eastern Sicily)	X	X		X	X	
Citadel, Ancient City and Fortress Buildings of Derbent			X	X		
Historic Quarter of the Seaport City of Valparaíso			X			
Jewish Quarter and St Procopius' Basilica in Třebíč		X	X			
Renaissance Monumental Ensembles of Úbeda and Baeza		X		X		
The White City of Tel-Aviv – the Modern Movement		X		X		
Dresden Elbe Valley		X	X	X	X	
Liverpool – Maritime Mercantile City		X	X	X		
Portuguese City of Mazagan (El Jadida)		X		X		

World Heritage city	(i)	(ii)	(iii)	(iv)	(v)	(vi)
Historic Centre of Macao		X	X	X		X
Historical Centre of the City of Yaroslavl		X		X		
Le Havre, the City Rebuilt by Auguste Perret		X		X		
Historic Centres of Berat and Gjirokastra, formerly Museum-City of Gjirokastra			X	X		
Old Bridge Area of the Old City of Mostar						X
Urban Historic Centre of Cienfuegos		X			X	
Harar Jugol, the Fortified Historic Town		X	X	X	X	
Old Town of Regensburg with Stadtamhof		X	X	X		
Bordeaux, Port of the Moon		X		X		
Old Town of Corfu				X		
Historic Centre of Camagüey				X	X	
Berlin Modernism Housing Estates		X		X		
Mantua and Sabbioneta		X	X			
Melaka and George Town, Historic Cities of the Straits of Malacca		X	X	X		
Protective town of San Miguel and the Sanctuary of Jesús Nazareno de Atotonilco		X		X		
San Marino Historic Centre and Mount Titano			X			
Cidade Velha, Historic Centre of Ribeira Grande		X	X			X
La Chaux-de-Fonds/Le Locle, Watchmaking Town Planning				X		
Episcopal City of Albi				X	X	
Seventeenth-Century Canal Ring Area of Amsterdam inside the Singelgracht	X	X		X		
Bridgetown and its Garrison		X	X	X		

Bibliography

Archival sources

National Board of Antiquities of Finland (NBA), Helsinki
Hbb1. 3, UNESCO, Maailmanperintöasiat 1972–1997.
Topografinen arkisto, Rauma IV: 1996–.

Old Rauma Renovation Centre (Vanhan Rauman korjausrakentamiskeskus) (OR Centre), Rauma
Vanhan Rauman asemakaavan toteuttamista ohjaava erityiselin (Old Rauma Committee) (VREe), Kokouspöytäkirjat 1982–2003.

Rauma City Archives (Rauman kaupunginarkisto) (RCA), Rauma
Kaupunginvaltuusto, kaupunginhallitus ja kaupunginkanslia (KV/KH/KK)
Erinäisten yleishallintoon liittyvien toimikuntien asiakirjat 1971–1976. Cfa: 12.
Vanhan Rauman suojelutoimikunnan 1973–1975.
Erinäisten kaavoitukseen ja yleisiin töihin liittyvien toimikuntien asiakirjat 1959–1970. Cff: 3.
Vanhan kaupunginosan saneeraustoimikunta 1961.
Keskustan asemakaavakilpailutoimikunta 1963–1965.
Erinäisten kaavoitukseen ja yleisiin töihin liittyvien toimikuntien ja työryhmien asiakirjat 1969–1975. Cff: 4.
Vanhankaupungin asemakaavatoimikunnan 1969–1972.
Kaupunginvaltuuston pöytäkirjat liitteineen 2004.
Neuvoa-antavien kokousten ja vastaavien pöytäkirjat ja muistiot liitteineen 1989–1991. CC4.
Hallinto-osasto (HO)
Keskustan asemakaavakilpailun aineisto, palkitut ja lunastetut ehdotukset 1964. Ha: 1.
Tekninen lautakunta / tekninen virasto (TEKLA)
Rakennus- ja kiinteistölautakunnan / teknisen lautakunnan pöytäkirjat liitteineen 1974–1981.

World Heritage Centre
World Heritage Committee & Bureau, Working Documents from 1976 to 1987.
World Heritage Sites inscribed in 1998 – Vol. 1 & 2.
World Heritage Sites inscribed in 1999 – Vol. 1 & 2.
World Heritage Sites, Nomination files (1978–1997), Africa and the Arab States.
World Heritage Sites, Nomination files (1978–1997), The Americas.

World Heritage Sites, Nomination files (1978–1997), Europe vol. I–II.
World Heritage Sites, Nomination files, (1978–1997), Latin America and the Caribbean.
World Heritage Sites, Nomination files (1978–1999), Rejected.

Databases

UNESCO World Heritage Centre, Documents, http://whc.unesco.org/en/documents/.
UNESCO World Heritage Centre, World Heritage List, http://whc.unesco.org/en/list.

Printed primary sources and web sources

Australia ICOMOS, *The Burra Charter. The Australia ICOMOS Charter for Places of Cultural Significance*. Victoria: Australia ICOMOS, 1999. Accessed October 15, 2015, http://australia.icomos.org/wp-content/uploads/BURRA_CHARTER.pdf.
Bosnia and Herzegovina, Nomination, "The Old City of Mostar," January 2005. Accessed January 15, 2013, http://whc.unesco.org/uploads/nominations/946rev.pdf.
Bürgerinitiative Pro Waldschlößchenbrücke. Accessed January 15, 2015, www.neue-waldschloesschenbruecke.de/unesco/unesco.htm.
Council of Europe Framework Convention on the Value of Cultural Heritage for Society, Faro 27.X.2005. Accessed October 15, 2015. www.coe.int/en/web/conventions/full-list/-/conventions/rms/0900001680083746.
El Harouny, Elisa, ed., *Suomalaisia puukaupunkeja. Hoito, kaavoitus ja suojelu*. Helsinki: Ympäristöministeriö, 1995.
Feilden, Bernard M. and Jokilehto, Jukka, *Management Guidelines for World Cultural Heritage Sites*. 2nd edn. Rome: ICCROM, 1998.
"From the Permanent Delegation of the Federal Republic of Germany for UNESCO to Francesco Bandarin, UNESCO-Recommendation on the Historic Urban Landscape, Paris, 17 December 2010." Accessed January 7, 2015, whc.unesco.org/document/117605.
Gazzola, Piero, "Foreword." In *The Monument for the Man: Records of the II International Congress of Restoration*, Venezia, 25–31 Maggio 1964. Padova: Marsilio, 1971. Accessed March 3, 2016, www.international.icomos.org/publications/hommeprein.pdf.
Helsingin Sanomat, October 25, 2004.
ICOMOS, Charter for the Conservation of Historic Towns and Urban Areas. ICOMOS: Washington, 1987. Accessed March 3, 2016, www.icomos.org/charters/towns_e.pdf.
ICOMOS, *Cultural Tourism*. International Scientific Symposium. ICOMOS, 10th General Assembly: Sri Lanka, 1993.
ICOMOS, *International Charter for the Conservation and Restoration of Monuments and Sites (The Venice Charter)*. Venice: The Second Congress of Architects and Specialists of Historic Buildings, 1964. Accessed February 15, 2016, www.icomos.org/charters/venice_e.pdf.
ICOMOS, The Role of ICOMOS in the World Heritage Convention. Accessed March 9, 2016, www.icomos.org/en/what-we-do/image-what-we-do/268-he-role-of-icomos-in-the-world-heritage-convention?showall=&limitstart=.
ICOMOS, *Threats to World Heritage Sites 1994–2004: An Analysis*, May 2005. Accessed June 12, 2014, www.icomos.org/world_heritage/Analysis%20of%20Threats%201994-2004%20final.pdf.

ICOMOS Austria, "The Wien-Mitte Project as Threat to the World Heritage Site 'Historic Centre of Vienna.'" In *ICOMOS World Report 2002–2003 on Monuments and Sites in Danger*. Accessed April 10, 2012, www.international.icomos.org/risk/2002/index.html.

ICOMOS News 8: 1, 1998.

International Relations and Security Network, ETH Zurich, "Hopeful rebirth for Bosnia's divided Mostar," February 3, 2004. Accessed June 3, 2041, www.isn.ethz.ch/isn/Current-Affairs/Security-Watch/Detail/?ots591=4888CAA0-B3DB-1461-98B9-E20E7B9C13D4&lng=en&id=107315.

Jokilehto, Jukka, Cameron, Christina, Parent, Michel and Petzet, Michael, *What is OUV? Defining the Outstanding Universal Value of Cultural World Heritage Properties* (2008). Monuments and sites XVI. Berlin: Hendrik Bäßler Verlag, 2008.

Jokilehto, Jukka, Cleere, Henry, Denyer, Susan and Petzet, Michael, *The World Heritage List: Filling the Gaps – an Action Plan for the Future*. München: ICOMOS, 2005.

KHO: 2006:2, 26.1.2006, 1236/1/05. Accessed April 21, 2012, www.kho.fi/paatokset/34553.htm.

LeBlanc, F., "An inside view of the Convention," *Monumentum*, Special Issue, 17–32, 1984.

Linnala, Minna and Nummelin, Liisa, *Toiveet ja todellisuus. Satakunnan rakennusperinnön hoito-projekti 1998–2000*. Pori: Satakunnan Museo, 2000.

Länsi-Suomi 1960–2007.

Modern heritage properties on the World Heritage List (as at July 2006). Accessed April 8, 2015, http://whc.unesco.org/uploads/activities/documents/activity-38-2.pdf.

Museum III: 1, 1950.

Organization of World Heritage Cities. Accessed December 15, 2015, www.ovpm.org/.

Partnerships for World Heritage Cities: Culture as a Vector for Sustainable Urban Development. World Heritage Papers 9. Paris: World Heritage Centre, 2003.

Policy for the Implementation of the ICOMOS World Heritage Mandate. ICOMOS Executive Committee 17 January 2006 and as amended in November 2007, in October 2010 and in October 2012. Accessed October 15, 2015, www.international.icomos.org/world_heritage/ICOMOS_WH_Policy_paper_201011_EN.pdf

Pressouyre, Léon, *The World Heritage Convention, Twenty Years Later*. Paris: UNESCO, 1996.

Rauman kaupungin kunnalliskertomus 1982.

Rauman kaupunki, kaavoitus, *Leikari-Lampolan asemakaavan muutos* (AK 01-084), 8.9.2004.

Rauman kaupunki, kaavoitus, *Rauman keskustan osayleiskaava*. Osayleiskaavan selostus, joka koskee 16.1.2003 päivättyä osayleiskaavakarttaa.

Rauman kaupunki, kaavoitus, *Rauman yleiskaava* 2.7.2003.

Report on the buffer zone of Old Rauma World Heritage site. National Board of Antiquities, Finland 23.1.2009. In the possession of Margaretha Ehrström.

Silva, Ronald, "Report of the President of ICOMOS, 1990–1995." *Thirty Years of ICOMOS. Scientific Journal* 5 (1995): 117–133.

Tabet, Jade, *Review of ICOMOS' working methods and procedures for the evaluation of cultural and mixed properties*. ICOMOS: Paris, 2010. Accessed October 15, 2015, www.icomos.org/world_heritage/WH_Committee_34th_session_Brasilia/JT_Final_report_en.pdf.

Tammivaara, A. B., "Rauma – piirteitä sen kehityksestä ja uusista suunnitteluista." In *Rauma. Retkeilijä*. No. 5. Helsinki: Finnish Literature Society, 1938, 76.

Tentative Lists, Austria, Great Spas of Europe. Accessed November 3, 2015, http://whc.unesco.org/en/tentativelists/5930/.

Tentative Lists, United Kingdom of Great Britain and Northern Ireland. Accessed November 3, 2015, http://whc.unesco.org/en/tentativelists/state=gb.

UNESCO, *Convention Concerning the Protection of World Cultural and Natural Heritage (The World Heritage Convention)*. Paris: UNESCO, 1972. Accessed February 15, 2016, http://whc.unesco.org/en/conventiontext/.

UNESCO, *Convention for the Safeguarding of Intangible Cultural Heritage*. Paris: UNESCO, October 17, 2003. Accessed July 12, 2015, www.unesco.org/culture/ich/en/convention.

UNESCO, Encouraging transmission of ICH: Living Human Treasures. Accessed December 15, 2015, www.unesco.org/culture/ich/index.php?pg=00061.

UNESCO, *Operational Guidelines for the implementation of the World Heritage Convention* Intergovernmental Committee for the Protection of the World Cultural and Natural Heritage. 20 October 1977. Paris: UNESCO. Accessed October 12, 2015, http://whc.unesco.org/archive/opguide77b.pdf.

UNESCO, *Operational Guidelines for the implementation of the World Heritage Convention*, October 1980. Intergovernmental Committee for the Protection of the World Cultural and Natural Heritage. WHC/2/Revised. Paris: UNESCO. Accessed October 12, 2015, http://whc.unesco.org/archive/opguide80.pdf.

UNESCO, *Operational Guidelines for the implementation of the World Heritage Convention*, January 1987. Intergovernmental Committee for the Protection of the World Cultural and Natural Heritage. WHC/2/Revised. Paris: UNESCO. Accessed October 12, 2015, http://whc.unesco.org/archive/opguide87.pdf.

UNESCO, *Operational Guidelines for the implementation of the World Heritage Convention*, 27 March 1992. Intergovernmental Committee for the Protection of the World Cultural and Natural Heritage. WHC/2/Revised. Paris: UNESCO. Accessed October 12, 2015, http://whc.unesco.org/archive/opguide92.pdf.

UNESCO, *Operational Guidelines for the implementation of the World Heritage Convention*, February 1994. Intergovernmental Committee for the Protection of the World Cultural and Natural Heritage. WHC/2/Revised. Paris: UNESCO. Accessed October 12, 2015, http://whc.unesco.org/archive/opguide94.pdf.

UNESCO, *Operational Guidelines for the implementation of the World Heritage Convention*, February 1996. Intergovernmental Committee for the Protection of the World Cultural and Natural Heritage. WHC96/2. Paris: UNESCO. Accessed October 12, 2015, http://whc.unesco.org/archive/opguide96.pdf.

UNESCO, *Operational Guidelines for the implementation of the World Heritage Convention*, 2 February 2005. Intergovernmental Committee for the Protection of the World Cultural and Natural Heritage. WHC-05/2. Paris: UNESCO. Accessed October 12, 2015, http://whc.unesco.org/archive/opguide05-en.pdf.

UNESCO, *Operational Guidelines for the implementation of the World Heritage Convention*, November 2011. Intergovernmental Committee for the Protection of the World Cultural and Natural Heritage. WHC-11/01. Paris: UNESCO. Accessed October 12, 2015, http://whc.unesco.org/archive/opguide11-en.pdf

UNESCO, *Operational Guidelines for the Implementation of the World Heritage Convention*, 8 July 2015. Intergovernmental Committee for the Protection of the World Cultural and Natural Heritage. WHC-15/01. Paris: UNESCO. Accessed October 12, 2015, http://whc.unesco.org/en/guidelines/.

UNESCO, The Proclamation of Masterpieces of the Oral and Intangible Heritage of Humanity. Accessed December 15, 2015, www.unesco.org/bpi/intangible_heritage/backgrounde.htm.

216   *Bibliography*

UNESCO, *Recommendation concerning Safeguarding and Contemporary Role of Historic Areas*. Paris UNESCO, 1976. Accessed March 3, 2016, http://portal.unesco.org/en/ev.php-URL_ID=13133&URL_DO=DO_TOPIC&URL_SECTION=201.html.

UNESCO, *Recommendation on the Historic Urban Landscape*. Paris: UNESCO, 2011. Accessed March 3, 2016, http://portal.unesco.org/en/ev.php-URL_ID=48857&URL_DO=DO_TOPIC&URL_SECTION=201.html.

UNESCO, State of Conservation of World Heritage Properties in Europe, Section II, Finland, Old Rauma. Accessed February 24, 2012, http://whc.unesco.org/archive/periodicreporting/EUR/cycle01/section2/582-summary.pdf.

Uusi Aika, 2 August 1983.

Uusi Rauma, 27 April 1994.

Valtioneuvosto, Valtioneuvoston päätös valtakunnallisista alueidenkäyttötavoitteista, November 30, 2000, Helsinki. Accessed 7 May, 2012, www.ymparisto.fi/download.asp?contentid=94382&lan=fi.

Xi'an Declaration on the Conservation of the Setting of Heritage Structures, Sites and Areas. Xi'an, China: 15th General Assembly of ICOMOS, 2005. Accessed February 15, 2016, www.icomos.org/charters/xian-declaration.pdf.

Printed secondary sources

Aa, Bart J.M. van der, *Preserving the Heritage of Humanity? Obtaining World Heritage Status and the Impacts of Listing*. Unpublished Ph.D. Thesis, University of Groeningen, 2005.

Aa, Bart J. M. van der, Groote, Peter D. and Huigen, Paulus P. P., "World Heritage as NIMBY? The Case of the Dutch Part of the Wadden Sea." In *The Politics of World Heritage: Negotiating Tourism and Conservation*, edited by David Harrison and Michael Hitchcock, 11–22. Clevedon: Channel View Publications, 2005.

Adams, Kathleen, "The Politics of Heritage in Tana Toraja, Indonesia: Interplaying the Local and the Global." *Indonesia and the Malay World* 31: 89 (2003): 91–107.

Affolder, Natasha, "Democratising or Demonising the World Heritage Convention?" *Victoria University of Wellington Law Review* 38: 2 (2007): 341–362.

Aitchison, Cara, "Heritage and Nationalism: Gender and the Performance of Power." In *Leisure/Tourism Geographies: Practices and Geographical Knowledge*, edited by David Crouch, 59–73. London: Routledge, 1999.

Araoz, Gustavo, "Protecting Heritage Places Under the New Heritage Paradigm & Defining its Tolerance for Change: a Leadership Challenge for ICOMOS. Message presented to ICOMOS Executive and Advisory committees in La Valletta, Malta, October 2009." Accessed April 24, 2012, www.fondazione-delbianco.org/seminari/progetti_prof/progview_PL.asp?start=1&idprog=283.

Ashworth, G. J. and Aa, Bart J. M. van der, "Strategy and Policy for the World Heritage Convention: Goals, Practices and Future Solutions." In *Managing World Heritage Sites*, edited by Anna Leask and Alan Fyall, 147–158. Amsterdam: Elsevier, 2006.

Ashworth, G. J., "The Conserved European City as Cultural Symbol: The Meaning of the Text." In *Modern Europe: Place, Culture and Identity*, edited by Brian Graham, 261–286. London: Arnold, 1998.

Ashworth, G. J. and Tunbridge J. E., *The Tourist-historic City: Retrospect and Prospect of Managing the Heritage City*. Amsterdam: Pergamon, 2000.

Ashworth G. J. and Graham, B., "Senses of Place, Senses of Time and Heritage." In *Senses of Place, Senses of Time*, edited by G. J. Ashworth and B. Graham, 3–12. Aldershot: Ashgate, 2005.

Askew, Marc, "The Magic List of Global Status. UNESCO, World Heritage and the Agendas of States." In *Heritage and Globalisation*, edited by Sophia Labadi and Colin Long, 19–44. Routledge: London, 2010.

Avarami, Erica, Mason, Randall and de la Torre, Marta, *Values and Heritage Conservation: Research Report*. Los Angeles, CA: Getty Conservation Institute, 2000. Accessed February 24, 2016, http://hdl.handle.net/10020/gci_pubs/values_heritage_research_report.

Bandarin, Francesco and Labadi, Sophia, *World Heritage: Challenges for the Millennium*. Paris: World Heritage Centre, 2007.

Beck, Ulrich, *The Cosmopolitan Vision*. Cambridge: Polity Press, 2006.

Beck, Wendy, "Narratives of World Heritage in Travel Guidebooks." *International Journal of Heritage Studies* 12: 6 (2006): 521–535.

Benevolo, Leonardo, "The City as an Expression of Culture: The Case of 14th Century Urbino." In *Partnerships for World Heritage Cities: Culture as a Vector for Sustainable Urban Development*, 17–20. World Heritage Papers 9. Paris: World Heritage Centre, 2003.

Bennett, Tony, *The Birth of the Museum: History, Theory, Politics*. London: Routledge, 1995.

Berkowitz, C. and Hoffmann, J., "The White City of Tel-Aviv." In *World Heritage and Buffer Zones*. World Heritage Papers 25, edited by Oliver Martin and Giovanna Piatti, 125–128. Paris: World Heritage Centre, 2007.

Berleant, Arnold, *Art and Engagement*. Philadelphia: Temple University Press, 1991.

Bonsdorff, Pauline von, *The Human Habitat: Aesthetic and Axiological Perspectives*. Lahti: International Institute of Applied Aesthetics, 1998.

Borsay, Peter, *The Image of Georgian Bath, 1700–2000: Towns, Heritage, and History*. Oxford: Oxford University Press, 2000.

Bortolotto, Chiara, "Globalising Intangible Cultural Heritage? Between International Arenas and Local Appropriations." In *Heritage and Globalisation*, edited by Sophia Labadi and Colin Long, 97–114. Routledge: London, 2010.

Byrne, Denis, "Western Hegemony in Archaeological Heritage Management." *History and Anthropology* 5: 2 (1991): 269–276.

Cameron, Christina, "Authenticity and the World Heritage Convention." In *Nara Conference on Authenticity: Proceedings*, edited by Knut Einar Larsen, 283–285. Paris: World Heritage Centre, 1995.

Cameron, Christina and Rössler, Mechtild, *Many Voices, One Vision: The Early Years of the World Heritage Convention*. Farnham, Surrey: Ashgate, 2013.

Carman, J., *Against Cultural Property: Archaeology, Heritage and Ownership*. London: Duckworth, 2005.

Choay, Françoise, *The Invention of the Historic Monument*. Cambridge: Cambridge University Press, 2001.

Clark, Peter, ed., *Cities in World History*. Oxford: Oxford University Press, 2013.

Clark, Peter, *European Cities and Towns 400–2000*. New York: Oxford University Press, 2009.

Cleere, Henry, "The Concept of 'Outstanding Universal Value' in the World Heritage Convention." *Conservation and Management of Archaeological Sites*, 1: 4 (1996): 227–233.

Cleere, Henry, "The Evaluation of Authenticity in the Context of the World Heritage Convention." In *Nara Conference on Authenticity: Proceedings*, edited by Knut Einar Larsen, 57–66. Paris: World Heritage Centre, 1995.

Cleere, Henry, "The Evaluation of Cultural Landscapes: The Role of ICOMOS." In *Cultural Landscapes of Universal Value*. Edited by Bernd von Droste, Harald Plachter and Mechtild Rössler, 50–59. Jena: Fischer, 1995.

Cleere, Henry, "The Uneasy Bedfellows: Universality and Cultural Heritage." In *Destruction and Conservation of Cultural Property*, edited by Robert Layton, Peter G. Stone and Julian Thomas, 22–29. London/New York: Routledge, 2001.

Cohen, Deborah and O'Connor, Maura, "Introduction: Comparative History, Cross-National History, Transnational History – Definitions." In *Comparison and History. Europe in Cross-National Perspective*, edited by Deborah Cohen and Maura O'Connor, ix–xxiv. New York: Routledge, 2004.

Creighton, Oliver, "Contested Townscapes: The Walled City as World Heritage." *World Archaeology* 39: 3 (2007): 339–354.

Cresswell, Tim, *Place. A Short Introduction*. Malden: Blackwell, 2004.

Crouch, David, "The Perpetual Performance and Emergence of Heritage." In *Culture, Heritage and Representation: Perspectives on Visuality and the Past*, edited by Emma Waterton and Steve Watson, 57–71. Surrey: Ashgate, 2010.

Crow, Graham and Allan, Graham, *Community Life. An Introduction to Local Social Relations*. New York: Harvester Wheatsheaf, 1994.

Davison, Graeme, "Heritage: From Patrimony to Pastiche." In *The Heritage Reader*, edited by Graham Fairclough, Rodney Harrison, John H. Jameson Jr and John Schofield, 31–41. London: Routledge, 2008.

Denslagen, Wim, *Romantic Modernism: Nostalgia in the World of Conservation*. Amsterdam: Amsterdam University Press, 2009.

Dicks, Bella, *Culture on Display: the Production of Contemporary Visitability*. Buckingham: Open University Press, 2003.

Diefendorf, Jeffry M. and Ward, Janet, "Introduction: Transnationalism and the German City." In *Transnationalism and the German City*, edited by Jeffry M. Diefendorf and Janet Ward, 1–10. New York: Palgrave Macmillan, 2014.

Dingli, Sandra M., "A Plea for Responsibility towards the Common Heritage of Mankind." In *The Ethics of Archaeology: Philosophical Perspectives on Archaeological Practice*, edited by Chris Scarre and Geoffrey Scarre, 219–241. Cambridge: Cambridge University Press, 2006.

Doyle, Barry, "A Decade of Urban History: Ashgate's Historical Urban Studies Series." *Urban History* 36: 3 (2009): 498–512.

Droste, Bernd von and Bertilsson, Ulf, "Authenticity and World Heritage." In *Nara Conference on Authenticity: Proceedings*, edited by Knut Einar Larsen, 3–15. Paris, World Heritage Centre, 1995.

Dyos, H. J., "Editorial." *Urban History Yearbook* (1974): 3–10.

El Harouny, Elisa, *Historiallinen puukaupunki suojelukohteena ja elinympäristönä: esimerkkeinä Vanha Porvoo ja Vanha Raahe*. Oulu: Oulun yliopisto, 2008.

El Harouny, Elisa, "Historiallinen puukaupunki – tulevaisuuden elävää kaupunkiympäristöä?" In *Eletty ja muistettu tila*, edited by Taina Syrjämaa and Janne Tunturi, 257–276. Helsinki: Finnish Literature Society, 2002.

Evans, Graeme, "Living in a World Heritage City: Stakeholders in the Dialectic of the Universal and Particular." *International Journal of Heritage Studies* 8: 2 (2002): 117–136.

Fairclough, Norman, *Analysing Discourse: Textual Analysis for Social Research*. London: Routledge, 2003.

Fairclough, Norman, "Discourse Analysis in Organizational Studies: The Case for Critical Realism." *Organization Studies* 26 (2005): 915–939.

Fowler, Peter, *World Heritage Cultural Landscapes 1992–2002*. World Heritage Papers 6. Paris: UNESCO World Heritage Centre, 2003.

Fowler, Peter, "World Heritage Cultural Landscapes, 1992–2002: A Review and Prospect." In *Cultural Landscapes: The Challenges of Conservation*, 16–28. World Heritage Papers 7. Paris: UNESCO / World Heritage Centre, 2003.

Fuge, Janina, "The Local, the National – and the Transnational? Spatial Dimensions in Hamburg's Memory of World War I during the Weimar Republic." In *Transnationalism and the German City*, edited by Jeffry M. Diefendorf and Janet Ward, 173–185. New York: Palgrave Macmillan, 2014.

Galland, Pierre, Lisitzin, Katri, Oudaille-Diethardt, Anatole and Young, Christopher, *World Heritage in Europe Today*. Paris: UNESCO, 2016.

Getty Conservation Institute, "Principles, Practice, and Process: A Discussion about Heritage Charters and Conventions." *The Getty Conservation Institute Newsletter* 19: 2 (2004), www.getty.edu/conservation/publications_resources/newsletters/19_2/dialogue.html.

Gibson, Lisanne, "Cultural Landscapes and Identity." In *Valuing Historic Environments*, edited by Lisanne Gibson and John Pendlebury, 67–92. Surrey: Ashgate, 2009.

Gibson, Lisanne and Pendlebury, John, "Introduction: Valuing Historic Environments." In *Valuing Historic Environments*, edited by Lisanne Gibson and John Pendlebury, 1–16. Surrey: Ashgate, 2009.

Graham, Brian, Ashworth, G. J. and Tunbridge, J. E., *A Geography of Heritage. Power, Culture and Economy*. London: Arnold, 2000.

Green, Carina, *Managing Laponia: A World Heritage as Arena for Sami Ethno-Politics in Sweden*. Uppsala: Uppsala University, 2009.

Gunn, Simon, "The Spatial Turn: Changing Histories of Space and Place." In *Identities in Space: Contested Terrains in the Western City since 1850*, edited by Simon Gunn and Robert J. Morris, 1–14. Aldershot: Ashgate, 2001.

Gunzburger Makaš, Emily, "Rebuilding Mostar: International and Local Visions of a Contested City and its Heritage." In *On Location: Heritage Cities and Sites*, edited by D. F. Ruggles, 151–168. New York: Springer, 2012.

Gunzburger Makaš, Emily, *Representing Competing Identities: Building and Rebuilding in Postwar Mostar, Bosnia-Herzegovina*. Unpublished Ph.D. Dissertation, Cornell University, 2007.

Hagen, Joshua, *Preservation, Tourism and Nationalism. The Jewel of the German Past*. Adlershot: Ashgate, 2006.

Hall, Melanie, "Introduction: Towards World Heritage." In *Towards World Heritage. International Origins of the Preservation Movement 1870–1930*, edited by Melanie Hall, 1–19. Aldershot: Ashgate, 2011.

Hall, Michael C., "Implementing the World Heritage Convention: What Happens after Listing?" In *Managing World Heritage Sites*, edited by Anna Leask and Alan Fyall, 22–30. Amsterdam: Elsevier, 2006.

Hall, Stuart, "Whose Heritage? Un-settling 'the Heritage,' Re-imagining the Post-nation." In *The Politics of Heritage: The Legacies of "Race,"* edited by Jo Littler and Roshi Naidoo, 23–35. London: Routledge, 2005.

Hardy, Matthew, ed., *The Venice Charter Revisited: Modernism, Conservation and Tradition in the 21st Century*. Newcastle upon Tyne: Cambridge Scholars Publishing, 2011.

Harrison, David, "Introduction." In *The Politics of World Heritage: Negotiating Tourism and Conservation*, edited by David Harrison and Michael Hitchcock, 1–10. Clevedon: Channel View Publications, 2005.

Harrison, Rodney, "The Politics of the Past. Conflict in the Use of Heritage in the Modern World." In *The Heritage Reader*, edited by Graham Fairclough, Rodney Harrison, John H. Jameson and John Schofield, 177–190. London: Routledge, 2008.

Harrison, Rodney, *Heritage: Critical Approaches*. London: Routledge, 2013.

Harvey, D. C., "A History of Heritage." In *Ashgate Research Companion to Heritage and Identity*, edited by Brian Graham and Perter Howard, 19–36. Abingdon: Ashgate, 2012.

Hayden, Dolores, "The Power of Place Project: Claiming Women's History in the Urban Landscape." In *Restoring Women's History through Historic Preservation*, edited by Gail Lee Dubrov and Jennifer B. Goodman, 199–213. Baltimore: The Johns Hopkins University Press, 2003.

Heino, Ulla, *Rauma – Idylliä ja tehokkuutta 1875–2000*. Rauma: Rauman kaupunki, 2002.

Hevia, James L., "World Heritage, National Culture, and the Restoration of Chengde." *Positions* 9: 1 (2001): 219–243.

Hitchcock, Michael, "Zanzibar Stone Town Joins the Imagined Community of World Heritage Sites." *International Journal of Heritage Studies* 8: 2 (2002): 153–166.

Hobsbawn, Eric, "Introduction: Inventing Traditions." In *The Invention of Tradition*, edited by Eric Hobsbawn and Terence Ranger, 1–14. Cambridge: Cambridge University Press, 1983.

Howard, Peter, "Historic Landscapes and the Recent Past: Whose History?" In *Valuing Historic Environments*, edited by Lisanne Gibson and John Pendlebury, 51–63. Surrey: Ashgate, 2009.

Häyrynen, Maunu, *Kuvitettu maa. Suomen kansallisen maisemakuvaston rakentuminen*. Helsinki: Finnish Literature Society, 2005.

Ito, Nobuo, "'Authenticity' Inherent in Cultural Heritage in Asia and Japan." In *Nara Conference on Authenticity: Proceedings*, edited by Knut Einar Larsen, 35–56. Paris, World Heritage Centre, 1995.

Jääskeläinen, Lauri, "KHO:n ja hallinto-oikeuksien päätöksiä." *Rakennettu Ympäristö* 2 (2006), 70. Accessed January 12, 2016, www.rakennustieto.fi/lehdet/ry/index/lehti/P_249.html.

Jacques, D. and Fowler, P., "Conservation of Landscapes in Post-industrial Countries." In *Cultural Landscapes of Universal Value – Components of a Global Strategy*, edited by B. von Droste, H. Plachter and M. Rössler, 412–419. Jena: Gustav Fischer Verlag, 1995.

Johnston, Chris, *What Is Social Value?* Canberra: Australian Heritage Commission, 1994.

Jokilehto, Jukka, "Authenticity: A General Framework for the Concept." In *Nara Conference on Authenticity: Proceedings*, edited by Knut Einar Larsen, 17–34. Paris: World Heritage Centre, 1995.

Jokilehto, Jukka, "Considerations on Authenticity and Integrity in World Heritage Context." *City & Time* 2: 1 (2006): 1–16. Accessed April 24, 2012, www.ct.ceci-br.org.

Jokilehto, Jukka, *A History of Architectural Conservation*. Oxford: ICCROM, 1999.

Jokilehto, Jukka, "Reflection on Historic Urban Landscapes as a Tool for Conservation." In *Managing Historic Cities*, edited by Ron van Oers and Sachiko Haraguchi, 53–63. World Heritage Papers 27. Paris: World Heritage Centre/UNESCO, 2010.

Jordan, Jennifer, *Structures of Memory: Understanding Urban Change in Berlin and Beyond*. Stanford: Stanford University Press, 2006.

Jumppanen, Seija, *Die Innere Differenzierung der Stadt Rauma*. Turku: Turun yliopisto, 1973.

Kenny, Nicholas and Madgin, Rebecca, "'Every Time I Describe a City': Urban History as Comparative and Transnational Perspective." In *Cities Beyond Borders: Comparative and Transnational Approaches to Urban History*, edited by Nicholas Kenny and Rebecca Madgin, 3–23. Farnham: Ashgate, 2015.

Kervanto Nevanlinna, Anja, "Classified Urban Spaces: Who Owns the History of Helsinki South Harbour?" In *Identities in Space. Contested Terrains in the Western City since 1850*, edited by S. Gunn and R. J. Morris, 19–38. London: Ashgate, 2001.

King, Anthony, "Terminologies and Types: Making Sense of Some Types of Dwellings and Cities." In *Ordering Space: Types in Architecture and Design*, edited by Karen A. Frank and Lynda H. Schneekloth, 127–146. New York: Van Nostrand Reinhold, 1994.

Kirshenblatt-Gimblett, Barbara, *Destination Culture. Tourism, Museums, and Heritage.* Berkeley: University of California Press, 1998.

Kirshenblatt-Gimblett, Barbara, "Intangible Heritage as Metacultural Production." *Museum International* 56: 1–2 (2004): 52–64.

Klimpke, Ute and Kammeier, H. Detlef, "Quedlinburg – 10 Years on the World Heritage List: East-West Transformations of a Small Historic Town in Central Germany." *International Journal of Heritage Studies*, 12: 2 (2006): 139–158.

Klusáková, Luda, "Between Urban and Rural Culture: Public Use of History and Cultural Heritage in Building Collective Identities (1990–2007)." In *Being a Historian – Opportunities and Responsibilities – Past and Present*, edited by Sven Mörsdorf, 195–218. A CLIOHRES-ISHA Reader II, www.cliohworld.net/docs/isha2_p.pdf.

Koivula, Jukka, "Vanha Rauma – Suojelukaavan jälkihoito." In *Rakennussuojelu*, edited by Jussi Rautsi and Kaija Santaholma, 128–129. Helsinki: Ympäristöministeriön Kaavoitus- ja rakennusosasto, 1987.

Koivula, Jukka, Nurmi-Nielsen, Anna, Saarinen, Kalle and Tyllilä, Irma, *Vanha Rauma – Old Rauma*. Rauma: Rauman museo, 1992.

Koponen, Olli-Paavo, *Täydennysrakentaminen. Arkkitehtuuri, historia ja paikan erityisyys.* Tampere: Tampere University of Technology, 2006.

Koshar, Rudy, *Germany's Transient Pasts. Preservation and National Memory in the Twentieth Century*. Chapel Hill: The University of North Carolina Press, 1998.

Kress, Gunther and van Leeuwen, Theo, *Reading Images: The Grammar of Visual Design*. London: Routledge, 2006.

Labadi, Sophia, "Representations of the Nation and Cultural Diversity in Discourses on World Heritage." *Journal of Social Archaeology* 7: 2 (2007): 147–170.

Labadi, Sophia, "A Review of the Global Strategy for a Balanced, Representative and Credible World Heritage List 1994–2004." *Conservation and Management of Archaeological Sites* 7:2 (2005): 89–102.

Labadi, Sophia, *UNESCO, Cultural Heritage, and Outstanding Universal Value. Value-based Analyses of the World Heritage and Intangible Cultural Heritage Conventions.* Lanham: AltaMira Press, 2013.

Labadi, Sophia, "World Heritage, Authenticity and Post-authenticity. International and National Perspectives." In *Heritage and Globalisation*, edited by Sophia Labadi and Colin Long, 66–84. Routledge: London, 2010.

Labadi, Sophia and Logan, William, eds., *Urban Heritage, Development and Sustainability: International Frameworks, National and Local Governance*. London/New York, Routledge/Taylor & Francis Group, 2016.

Laenen, Marc, "Authenticity in Relation to Development." In *Nara Conference on Authenticity: Proceedings*, edited by Knut Einar Larsen, 351–357. Paris: World Heritage Centre, 1995.

Larkham, Peter, J., *Conservation and the City*. London: Routledge, 1996.

Larsen, Knut Einar, "Preface." In *Nara Conference on Authenticity: Proceedings*, edited by Knut Einar Larsen, xi–xiii. Paris: World Heritage Centre, 1995.

Layton, Robert and Titchen, Sarah, "Uluru: An Outstanding Australian Aboriginal Cultural Landscape." In *Cultural Landscapes of Universal Value*. Edited by Bernd von Droste, Harald Plachter and Mechtild Rössler, 174–181. Jena: Fischer, 1995.

Leask, Anna, "World Heritage Site Designation." In *Managing World Heritage Sites*, edited by Anna Leask and Alan Fyall, 5–19. Elsevier: Amsterdam, 2006.

Levebvre, Henri, *The Production of Space*. Translated by Donald Nicholson-Smith. Oxford: Blackwell Publishing, 1991.

Lipe, Walter, "Value and Meaning in Cultural Resources." In *Approaches to the Archaeological Heritage. A Comparative Study of World Cultural Resource Management Systems*, edited by Henry Cleere, 1–11. Cambridge: Cambridge University Press, 1984.

Littler, Jo, "Introduction: British Heritage and the Legacies of 'Race.'" In *The Politics of Heritage: The Legacies of "Race,"* edited by Jo Littler and Roshi Naidoo, 1–20. London: Routledge, 2005.

Logan, William S., "Globalizing Heritage: World Heritage as a Manifestation of Modernism and Challenges from the Periphery." In *20th Century Heritage: Our Recent Cultural Legacy: Proceedings of the Australia International Council for Monuments and Sites National Conference 2001*, edited by David Jones, 51–57. University of Adelaide & Australia ICOMOS: Adelaide, 2001.

Long, Colin and Labadi, Sophia, "Introduction." In *Heritage and Globalisation*, edited by Sophia Labadi and Colin Long, 1–16. London: Routledge, 2010.

Lowenthal, David, *The Heritage Crusade and the Spoils of History*. Cambridge: Cambridge University Press, 1998.

Lowenthal, David, *The Past is a Foreign Country*. Cambridge: Cambridge University Press, 1985.

MacCannell, Dean, *The Tourist: a New Theory of the Leisure Class*. Berkeley: University of California Press, 1999.

Macdonald, Sharon, *Difficult Heritage: Negotiating the Nazi Past in Nuremberg and Beyond*. London: Routledge, 2009.

Macdonald, Sharon, *Memorylands. Heritage and Identity in Europe Today*. New York: Routledge, 2013.

Maddern, Joanne, "Huddled Masses Yearning to Buy Postcards: The Politics of Producing Heritage at the Statue of Liberty–Ellis Island National Monument." *Current Issues in Tourism* 7: 4 (2004): 303–314.

Madgin, Rebecca, *Heritage, Culture and Conservation: Managing the Urban Renaissance*. Saarbrücken: Verlag Dr. Müller, 2009.

Mason, Randall, "Assessing Values in Conservation Planning: Methodological Issues and Choices." In *Assessing the Values of Cultural Heritage*, edited by Marta de la Torre, 5–30. Los Angeles: The Getty Conservation Institute, 2002.

Mazlish, Bruce, "An Introduction to Global History." In *Conceptualizing Global History*, edited by Bruze Mazlish and Ralph Buultjens, 1–24. Boulder: Westview Press, 1993.

Meskell, Lynn, "Negative Heritage and Past Mastering in Archaeology." *Anthropological Quarterly* 75: 3 (2002): 557–574.

Millar, Sue, "Stakeholders and Community Participation." In *Managing World Heritage Sites*, edited by Anna Leask and Alan Fyall, 37–54. Elsevier: Amsterdam, 2006.

Miller, Michael James Michael, *The Representation of Place: Urban Planning and Protest in France and Great Britain, 1950–1980*. Aldershot: Ashgate, 2003.

Munasinghe, Harsha, *Urban Conservation and City Life: Case Study of Port City of Galle*. Oulu: University of Oulu, 1998.

Nilsson Dahlström, Åsa, *Negotiating Wilderness in a Cultural Landscape: Predators and Saami Reindeer Herding in the Laponian World Heritage Area*. Uppsala: Uppsala University, 2003.

Oksanen, Annukka, *Paikallisuuden ja kansainvälisyyden kohtaaminen luonnonsuojelussa. Tapaustutkimuksena Natura 2000 -ympäristökonflikti Lounais-Suomessa*. Turku: Turun yliopisto, 2003.

Oers, Ron, van, "Managing Cities and the Historic Urban Landscape Initiative – An Introduction." In *Managing Historic Cities*, edited by Ron van Oers and Sachiko Haraguchi, 7–17. World Heritage Papers 27. Paris: World Heritage Centre/UNESCO, 2010.

Oers, Ron, van, "Preventing the Goose with the Golden Eggs from Catching Bird Flu – UNESCO's Efforts in Safeguarding the Historic Urban Landscape." Keynote paper for the 42nd Congress of the International Society of City and Regional Planners (ISoCaRP). *Cities between Integration and Disintegration*, 14–18 September, 2006, Istanbul, Turkey.

Oers, Ron, van and Haraguchi, Sachiko eds., *Managing Historic Cities*. World Heritage Papers 27 Paris: World Heritage Centre/UNESCO, 2010.

Omland, Atle, "The Ethics of the World Heritage Concept." In *The Ethics of Archaeology: Philosophical Perspectives on Archaeological Practice*, edited by Chris Scarre and Geoffrey Scarre, 242–259. Cambridge: Cambridge University Press, 2006.

Otter, Chris, *The Victorian Eye: A Political History of Light and Vision in Britain, 1800–1910.* Chicago: University of Chicago Press, 2008.

Paasi, Anssi, "Rajat ja identiteetti globalisoituvassa maailmassa." In *Eletty ja muistettu tila*, edited by Janne Tunturi and Taina Syrjämaa, 154–176. Helsinki: Finnish Literature Society, 2002.

Pendlebury, John, Townshend, Tim and Gilroy, Rose, "Social Housing as Heritage: The Case of Byker, Newcastle upon Tyne." In *Valuing Historic Environments*, edited by Lisanne Gibson and John Pendlebury, 179–200. Surrey: Ashgate, 2009.

Petzet, Michael, "Conservation or Managing Change." Paper presented in *Conservation Turn – Return to Conservation: Challenges and Chances in a Changing World* (Prague, 5–9 May 2010). Accessed April 24, 2012, www.icomos.de/.

Petzet, Michael, "Introduction," in *What Is OUV? Defining the Outstanding Universal Value of Cultural World Heritage Properties*, edited by Jukka Jokilehto et al., 7–10. Monuments and sites XVI. Berlin: Hendrik Bäßler Verlag, 2008.

Petzet, Michael, "'In the Full Richness of Their Authenticity' – The Test of Authenticity and the New Cult of Monuments." In *Nara Conference on Authenticity: Proceedings*, edited by Knut Einar Larsen, 85–99. Paris: World Heritage Centre, 1995.

Poulios, Ioannis, "Moving Beyond a Values-based Approach to Heritage Conservation." *Conservation and Management of Archaeological Sites* 12: 2 (2010): 170–188.

Rao, Kishore, "A New Paradigm for the Identification, Nomination and Inscription of Properties on the World Heritage List." *International Journal of Heritage Studies* 16: 3 (2010): 161–172.

Relph, Edward, *Place and Placelessness*. London: Pion, 1976.

Rico, Trinidad, "Negative Heritage: The Place of Conflict in World Heritage." *Conservation and Management of Archaeological Sites* 10: 4 (2008): 344–352.

Riegl, Alois, "The Modern Cult of Monuments: Its Character and its Origins." *Oppositions* 25 (1982) [1903]: 20–51.

Ringbeck, Birgitta, "Die Entscheidung zum Dresdener Elbtal. Ein Kurzbericht zur 33. Sitzung des UNESCO-Welterbekomitees vom 22. bis 30. Juni 2009 Sevilla, Spanien." Accessed December 15, 2012, www.unesco.de/3652.html.

Robertson, Roland, "Glogalization: Time-Space and Homogeneity-Heterogeneity." In *Global Modernities*, edited by M. Featherstone, S. Lash and R. Robertson, 25–44. London: Sage, 1995.

Rodger, Richard and Herbert, Joanna, eds., *Testimonies of the City: Identity, Community and Change in a Contemporary Urban World*. Aldershot: Ashgate, 2007.

Rodger, Richard and Sweet, Roey, "The Changing Nature of Urban History." *History in Focus*, February 2008. Accessed March 3, 2015, www.history.ac.uk/ihr/Focus/City/articles/sweet.html#t14.

Rodman, Margaret C., "Empowering Place: Multilocality and Multivocality." In *The Anthropology of Space and Place: Locating Culture*, edited by Setha M. Low and Denise Lawrence-Zúñiga, 204–223. Malden, MA: Blackwell, 2003.

Ronström, Owe, "Kulturarvspolitik. Vad skyltar kan berätta." In *Kritisk etnologi. Artiklar till Åke Daun*, edited by Barbro Blehr, 60–108. Stockholm: Prisma, 2000.

Rosenfeld, Gavriel D., *Munich and Memory: Architecture, Monuments, and the Legacy of the Third Reich*. Berkley: University of California Press, 2000.

Rössler, Mechtild, "Linking Nature and Culture: World Heritage Cultural Landscapes." In *Cultural Landscapes: the Challenges of Conservation*, 10–15. World Heritage Papers 7. Paris: UNESCO World Heritage Centre, 2003.

Sajo, Jim, "Sun Sets on U.S. Military's 60-Year Stay in Verona, Italy." *Stars and Stripes*, June 13, 2004. Accessed June 22, 2011, www.stripes.com/military-life/sun-sets-on-u-s-military-s-60-year-stay-in-verona-italy-1.31754.

Samuel, Raphael, *Theatres of Memory. Vol. 1. Past and present in contemporary culture*. London: Verso, 1994.

Scott, Julie, "World Heritage as a Model for Citizenship: The Case of Cyprus." *International Journal of Heritage Studies* 8: 2 (2002): 99–115.

Seppänen, Maaria, *Global Scale, Local Place? The Making of the Historic Centre of Lima into a World Heritage Site*. Helsinki: University of Helsinki, 1999.

Sierp, Aline, *History, Memory, and Trans-European Identity. Unifying Divisions*. New York: Routledge, 2014.

Singh, J. P., *United Nations Educational, Scientific and Cultural Organization (UNESCO). Creating Norms for a Complex World*. New York: Routledge, 2011.

Smith, Julian, "Marrying the Old with the New in Historic Urban Landscapes." In *Managing Historic Cities*, edited by Ron van Oers and Sachiko Haraguchi, 45–51. World Heritage Papers 27. Paris: World Heritage Centre/UNESCO, 2010.

Smith, Laurajane, "Deference and Humility: The Social Values of the Country House." In *Valuing Historic Environments*, edited by Lisanne Gibson and John Pendlebury, 33–50. Surrey: Ashgate, 2009.

Smith, Laurajane "Heritage, Gender and Identity." In *Ashgate Research Companion to Heritage and Identity*, edited by Brian Graham and Perter Howard, 159–178. Abingdon: Ashgate, 2012.

Smith, Laurajane, *Uses of Heritage*. New York: Routledge, 2006.

Smith, Melanie, "A Critical Evaluation of the Global Accolade: The Significance of World Heritage Site Status for Maritime Greenwich." *International Journal of Heritage Studies* 8: 2 (2002): 137–152.

Stovel, Herb, "Effective Use of Authenticity and Integrity as World Heritage Qualifying Conditions." *City & Time* 2: 3 (2007): 21–36, www.ceci-br.org/novo/revista/docs2007/CT-2007-71.pdf.

Stovel, Herb, "ICOMOS Position Paper." In *World Heritage and Buffer Zones. International Expert Meeting on World Heritage and Buffer Zones, Davos, Switzerland 11 – 14 March 2008*, edited by Oliver Martin and Giovanna Piatti, 23–42. Paris: UNESCO World Heritage Centre, 2009.

Swenson, Astrid, *The Rise of Heritage: Preserving the Past in France, Germany and England 1789–1914*. Cambridge: Cambridge University Press, 2013.

Titchen, Sarah M., "On the Construction of 'Outstanding Universal Value'. Some Comments on the Implementation of the 1972 UNESCO World Heritage Convention." *Conservation and Management of Archaeological Sites*, 1: 4 (1996): 235–242.

Titchen, Sarah M., *On the Construction of Outstanding Universal Value: UNESCO's World Heritage Convention and the Identification and Assessment of Cultural Places for Inclusion in the World Heritage List*. Unpublished PhD. diss., Australian National University, 1995.

Tsing, Anna, "The Global Situation." *Cultural Anthropology* 15: 3 (2000): 327–360.

Tunbridge, J. E. and Ashworth, G. J., *Dissonant Heritage: the Management of the Past as a Resource in Conflict*. Chichester: Wiley, 1996.

Tunbridge, John E., "Whose Heritage to Conserve? Cross-cultural Reflections on Political Dominance and Urban Heritage Conservation." In *The Heritage Reader*, edited by Graham Fairclough, Rodney Harrison, John H. Jameson and John Schofield, 235–244. London: Routledge, 2008. First published in 1984.

Tuomi-Nikula, Outi, "Alasaksalainen hallitalo kertoo pientä historiaa. Talo eurooppalaisen etnologian tutkimuskohteena Saksassa." In *Historioita ja historiallisia keskusteluja*, edited by Sami Louekari and Anna Sivula, 244–269. Turku: Turun historiallinen yhdistys, 2004.

Tuominen, Laura, "'Museoiminen' metaforana ja rakennussuojelukriittisenä argumenttina." In *Rakkaudesta kaupunkiin*, edited by Renja Suominen-Kokkonen, 87–98. Helsinki: Taidehistorian seura, 2004.

Turner, Michael, "UNESCO Recommendation on the Historic Urban Landscape." In *Understanding Heritage: Perspectives in Heritage Studies*, edited by Marie-Theres Albert, Roland Bernecker and Britta Rudolff, 77–87. Berlin: De Gruyter, 2013.

Turtinen, Jan, "Globalising Heritage – On UNESCO and the Transnational Construction of a World Heritage'." *SCORE Rapportserie* No. 12, 2000. Accessed February 2, 2012, www.score.su.se/polopoly_fs/1.26651.1320939806!/200012.pdf.

Urry, John, *The Tourist Gaze*. 2nd edn. Sage: London, 2002.

Vahtikari, Tanja, "From National to World Heritage via the Regional: Harmonizing Heritage in the Nordic Countries." In *UNESCO and World Heritage: National Contexts, International Dynamics*, edited by Casper Andersen and Irena Kozymka. Ashgate (forthcoming).

Vahtikari, Tanja, "Historic Cities, World Heritage Value and Change." In *Touring the Past: Uses of History in Tourism*, edited by Auvo Kostiainen and Taina Syrjämaa, 132–150. Savonlinna: Matkailualan verkostoyliopisto, 2008.

Vahtikari, Tanja, "Merkityksin rakennettu Vanha Rauma: suomalaisen historiallisen kaupungin varhainen määrittely valintoina ja vuoropuheluna 1900–1970." In *Muistin kaupunki. Tulkintoja kaupungista muistin ja muistamisen paikkana*, edited by Katri Lento and Pia Olsson (Helsinki: Finnish Literature Society, 2013).

Vahtikari, Tanja, "Miten Vanhasta Raumasta tuli maailmanperintökohde?" In *Kotina suojeltu talo: Arkea, elämää ja rakennussuojelua Suomessa ja Saksassa*, edited by Outi Tuomi-Nikula and Eeva Karhunen, 97–117. Pori: Kulttuurituotannon ja maisemantutkimuksen laitos, 2007.

Vahtikari, Tanja, "Urban Interpretations of World Heritage: Re-defining the City." In *Reclaiming the City: Innovation, Culture, Experience*, edited by Marjaana Niemi and Ville Vuolanto, 63–79. Helsinki: SKS, 2003.

Valderrama, Fernando (1995), *A History of UNESCO*. UNESCO: Paris.

Vuolteenaho, Jani and Berg, Lawrence D., "Towards Critical Toponymies." In *Critical Toponymies: The Contested Politics of Place Naming*, edited by Lawrence D. Berg and Jani Vuolteenaho, 1–18. Surrey: Ashgate, 2009.

Walton, John K. and Wood, Jason, "Reputation and Regeneration: History and the Heritage of the Recent Past in the Re-making of Blackpool." In *Valuing Historic Environments*, edited by Lisanne Gibson and John Pendlebury, 115–137. Surrey: Ashgate, 2009.

Warren, K. J., "A Philosophical Perspective on the Ethics and Resolution of Cultural Property Issues." In *The Ethics of Collecting Cultural Property: Whose Culture?*, edited by P. M. Messenger, 1–25. Albuquerque: University of New Mexico, 1999.

Waterton, Emma, "Branding the Past: The Visual Imagery of England's Heritage." In *Culture, Heritage and Representation: Perspectives on Visuality and the Past*, edited by Emma Waterton and Steve Watson, 155–172. Surrey: Ashgate, 2010.

Waterton, Emma and Smith, Laurajane, "The Recognition and Misrecognition of Community Jeritage." *International Journal of Heritage Studies*, 16: 1–2 (2010): 4–15.

Waterton, Emma, Smith, Laurajane and Campbell, Gary, "The Utility of Discourse Analysis to Heritage Studies: The Burra Charter and Social Inclusion." *International Journal of Heritage Studies* 12: 4 (2006): 339–355.

Watson, Steve and Waterton, Emma, "Introduction: A Visual Heritage." In *Culture, Heritage and Representation: Perspectives on Visuality and the Past*, edited by Emma Waterton and Steve Watson, 1–16. Surrey: Ashgate, 2010.

Whelan, Yvonne, "Mapping Meanings in the Cultural Landscape." In *Senses of Places: Senses of Time*, edited by G. J. Ashworth and Brian Graham, 61–71. Aldershot: Ashgate, 2005.

Winter, Tim, "Heritage Tourism: The Dawn of a New Era?" In *Heritage and Globalisation*, edited by Sophia Labadi and Colin Long, 117–129. Routledge: London, 2010.

Worthing, Derek and Bond, Stephen, *Managing Built Heritage: The Role of Cultural Significance*. Oxford: Blackwell Publishing, 2008.

Interviews

Below I list each interviewee's name, his or her status/relation to Old Rauma and the date of the interview/interviews. In the endnotes the interviewees are usually referred to by their status/relation to Old Rauma and the date of interview.

Olli Ajanko, Old Rauma property owner and former shopkeeper, 7.11.2002.

Mari Aspola, Old Rauma resident and Old Rauma Society active, 27.11.2002.

Margaretha Ehrström, national conservation authority, World Heritage coordinator in the National Board of Antiquities, 29.1.2002, 23.11.2010.

Harri Holmala, city solicitor, treasurer of the Old Rauma Foundation, 28.11.2002.

Toivo Jaakkola, local politician, member of Rauma city council (1957–1984) and city government (1985–1988), 27.11.2002.

Reino Joukamo, former city of Rauma town planning architect, 27.3.2002.

Pentti Koivu, mayor of Rauma (in 2002), 9.4.2002.

Jukka Koivula, architect, city planner and Old Rauma resident, 6.11.2002.

Hilkka Kuisma, Old Rauma resident and Old Rauma guide, 26.11.2002.

Sinikka Kuusava, Old Rauma resident, 25.10.2002.

Pekka Kärki, national conservation authority, head of the Department of Built Heritage in the National Board of Antiquities (in 2002), 21.2.2002.

Mauno Lehtonen, local politician, member of Rauma city council (1961–2008), chair of the city government (1977–1991), 25.10.2002.

Henrik Lilius, national conservation authority, head of National Board of Antiquities (in 2002), the chairperson of the World Heritage Committee (2001–2002), 9.12.2002.

Juha Mantere, Old Rauma shopkeeper and Old Rauma resident, 3.11.2003, 6.11.2003.

Maire Mattinen, national conservation authority, 4.6.2004.

Uljas Mäkelä, member of Parliament (1962–1978) and mayor of Rauma (1978–1987), 4.11.2003.

Anna Nurmi-Nielsen, local conservation authority, director of Rauma city museum and Old Rauma resident (in 2003), 5.7.2002, 24.11.2003.

Pekka Oivanen, head of Rauma Tourism Office, 26.11.2002.

Pekka Palmu, Old Rauma shopkeeper, 28.11.2002.

Aino Pohjanoksa, member of Parliament (1983–1991), and member of Rauma city council (1977–1997; chair 1993–1996), 6.11.2002.

Leena Rautavuori, Old Rauma resident and Old Rauma Society active, 7.11.2002.

Ulla Räihä, city of Rauma town planning architect, 6.11.2003.

Kalle Saarinen, local conservation authority/Old Rauma advisory architect and Old Rauma resident, 5.11.2003.

Irma Tyllilä, city of Rauma town planning architect (until 2005), 27.3.2002.

Kari Valo, Old Rauma shopkeeper, 4.11.2002.

Aatos Virtanen, local politician, chair of Rauma city council (1977–1992), 26.11.2002.

Index

Convention concerning the Protection of the World Cultural and Natural Heritage *see* World Heritage Convention
Convention for the Safeguarding of Intangible Cultural Heritage (2003 Convention) 50–2, 113
Corfu, Greece 66, 68
Coro, Venezuela 122
Council of Europe Framework Convention on the Value of Cultural Heritage for Society 31
criteria 6, 9, 11–12, 18–19, 22–4, 31–3, 37, 46–8, 51, 70–1, 74, 77, 83, 112–13, 120, 157, 186, 190–1: comparison of 26–7; criterion i 25, 28, 32, 101, 190; criterion ii 25, 32, 68, 191; criterion iii 25, 32, 70; criterion iv 25, 28, 32, 65, 70, 154–5, 191; criterion v 28, 32, 70, 80, 101–2, 113, 154–5, 190; criterion vi 19, 28, 32, 69–71, 87, 119, 190; evolvement of 25–8, 32; use of in reference to cities 204–11
Critical Discourse Analysis 10
Croatian Democratic Union 86
cultural landscape 7, 12, 44, 46–7, 52, 55, 80, 107–8, 119, 123, 131, 184, 190

Damascus, Syrian Arab Republic 101
'democratization of heritage,' trend of 53
Derbent, Russian Federation 75
Djenné, Mali 112, 127
Dresden, Germany 56, 63, 78, 80, 84, 102, 171, 173, 175, 187, 189; deletion from the World Heritage List 80

Edinburgh, UK 56
Ehrström, Margareta 163, 174
Elisabeth, Empress of Russia 81
Essaouira, Morocco 65, 88
Eurocentrism 6, 68, 100–1
Eyck, Jan van 70

Feilden, Bernard M. 29
Ferrara, Italy 82, 87
Florence, Italy 20, 65, 66

Gama, Vasco da 72
Garibaldi, Giuseppe 77
Gazzola, Piero 82
Gdańsk, Poland 75, 118, 125
General Assembly of States Parties 9, 18
gentrification 113, 144
George Town, Malaysia 68, 113, 114, 123

Ghadamés, Libyan Arab Jamahiriya 101, 122
Gjirokastra, Albania 87
Global Strategy for a Balanced, Credible and Representative World Heritage List (Global Strategy) 12, 23, 44–7, 52, 54–5, 63, 83, 99, 101–3, 108–10, 113–14, 118, 120
Global Study 45–6
"glocalism" 172
Goiás, Brazil 81, 104, 113
Graz, Austria 56
Grotius, Hugo 19
Guanajuato, Mexico 70, 81
Guimares, Portugal 65

Hague Convention 20
Harar Jugol, Ethiopia 98, 110, 116
heritage (social and cultural construction) 4–7
heritage values 1–2, 5, 24–5, 28–32, 48–9, 51, 98, 117, 119–20, 169–70, 174, 185, 190–1; aesthetic value 10, 25, 29–32, 104, 115, 147, 155; economic value 29; historical value 12, 29–30, 48, 62, 70, 77, 104, 118, 148, 152; intrinsic 2, 5, 12, 18, 23–4, 30, 38, 46, 120, 124, 185, 187; monumental value *see* monumental(ity); social value 25, 28–32, 81, 112–14, 116, 119–20, 191; spiritual value 30–1, 114; typology 29
Hidalgo, Miguel 72
Hiroshima Peace Memorial 84
historic cities 7, 13, 54, 56, 77, 79, 87, 100, 105–8, 111–13, 118, 120, 122, 126, 130, 145, 147, 152, 159, 160, 191
historic urban landscapes 56, 104–5, 107, 190

Ibiza, Spain 115
ICCROM (International Centre for the Study of the Preservation and Restoration of Cultural Property) 20, 36
ICOMOS (International Council on Monuments and Sites): advisory role 19–20; agency 1–2; evaluation documents 9–10, 23, 35–7; evaluations and the "authorized heritage discourse" 6, 187–8; expertise within World Heritage framework 32–8; national commissions 34–5; national nominations and 37, 49; negative statements 3, 7–9; objectivity of 24, 32–7, 184–5; scientific